Prefac

Singapore Math® Tests is a series of structured assessments to help teachers evaluate student progress. The tests align with the content of Primary Mathematics Common Core textbooks.

Each level offers differentiated tests (Test A and Test B) to suit individual needs. Tests consist of multiple-choice questions that assess comprehension of key concepts and free response questions that demonstrate problem solving skills. Three continual assessments cover topics from earlier units and a year-end assessment covers the entire curriculum.

Test A focuses on key concepts and fundamental problem solving skills.

Test B focuses on the application of analytical skills, thinking skills, and heuristics.

Singapore Math®

Tests

2B

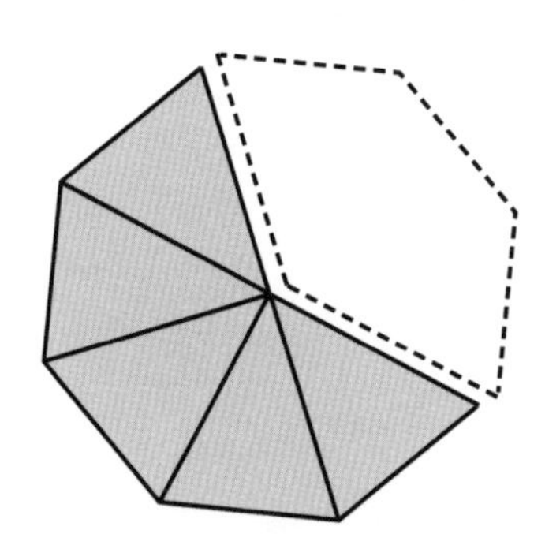

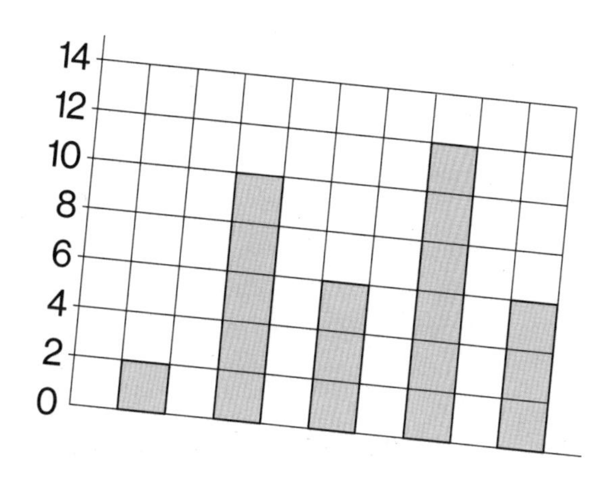

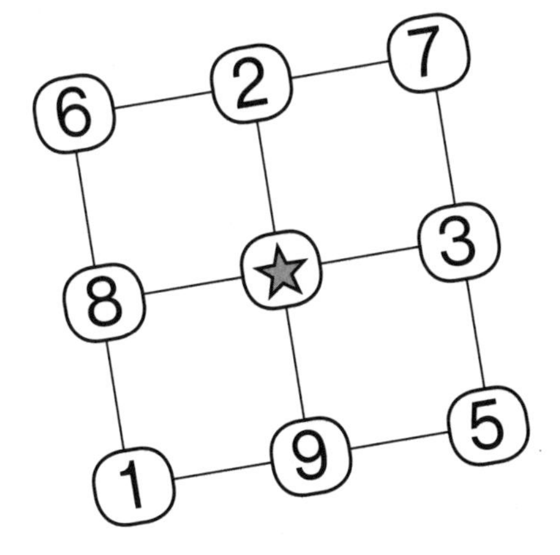

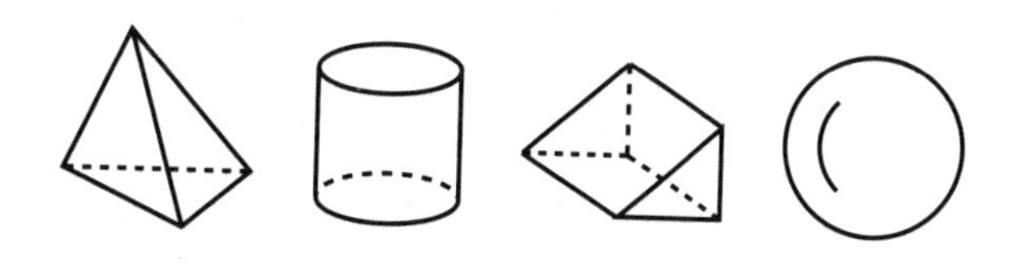

Differentiated Unit Tests
Continual Assessments

Singapore Math Inc.®

BLANK

Contents

BLANK

Name: ______________________ Date: __________

Test A

25 min

30

Score

Unit 6 Addition and Subtraction

Section A (2 points each)
Circle the correct option: **A**, **B**, **C**, or **D**.

1. What is the missing number in 14 + [?] = 60?

   A 56 **B** 46

   C 44 **D** 50

2. What is 20 less than 375?

   A 373 **B** 175

   C 355 **D** 395

3. What is the value of 62 + 15?

   A 77 **B** 72

   C 75 **D** 70

4. What is the value of 278 – 50?

A 273 **B** 228

C 218 **D** 220

5. What is 700 – 99?

A 600 **B** 691

C 610 **D** 601

Section B (2 points each)

6. $84 + 4 =$ ☐

7. $230 + 200 =$ ☐

8. $63 - 13 =$ ☐

9. $63 + 13 =$ ☐

10. $99 + 16 =$ ☐

11. $100 - 89 =$ ☐

12. Write the missing number.

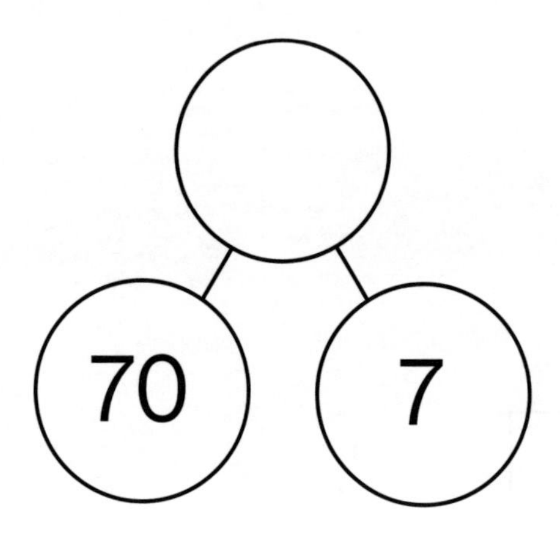

13. is 5 less than 456.

14. Find the value of 301 – 98.

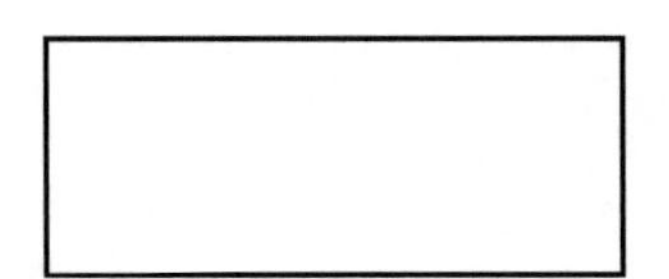

15. What number is 20 more than 637?

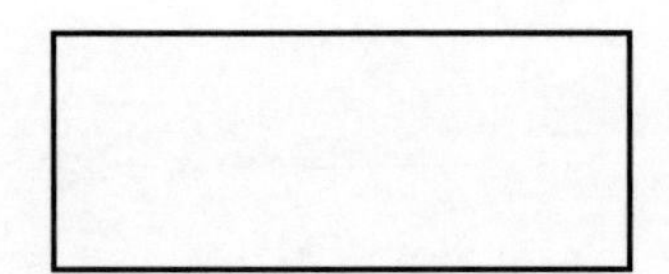

Name: ______________________ Date: __________

Test B

25 min

30

Score

Unit 6 Addition and Subtraction

Section A (2 points each)
Circle the correct option: **A**, **B**, **C**, or **D**.

1. What is the value of 39 + 25?

   A 54 **B** 64

   C 65 **D** 46

2. What is the missing number?
   5 tens and 10 ones make __?__.

   A 51 **B** 15

   C 105 **D** 60

3. Subtract 99 from 998.
   What is the answer?

   A 899 **B** 999

   C 900 **D** 908

4. What number is 10 more than 92?

A 93

B 103

C 102

D 82

5. How many tens do you need to subtract from 100 to make 50?

A 2

B 3

C 4

D 5

Section B (2 points each)

6. ☐ is 7 less than 706.

7. There are 78 apples in a box.

☐ more are needed to make 100.

8. Write **True** or **False**.

45 + 20 = 25

☐

9. What number is 1 less than 900?

10. Write the missing number.

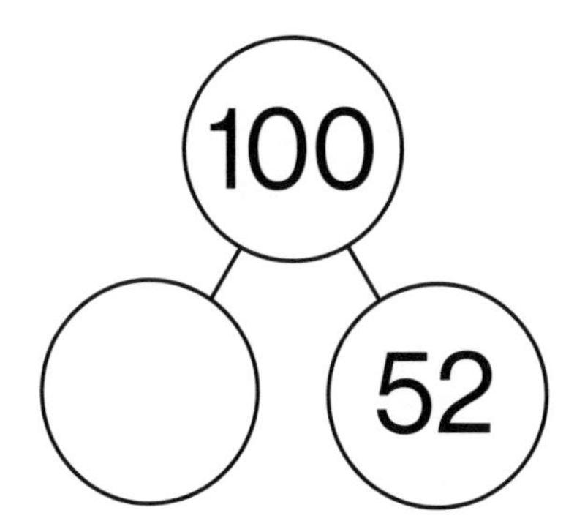

11. $299 + 89 =$

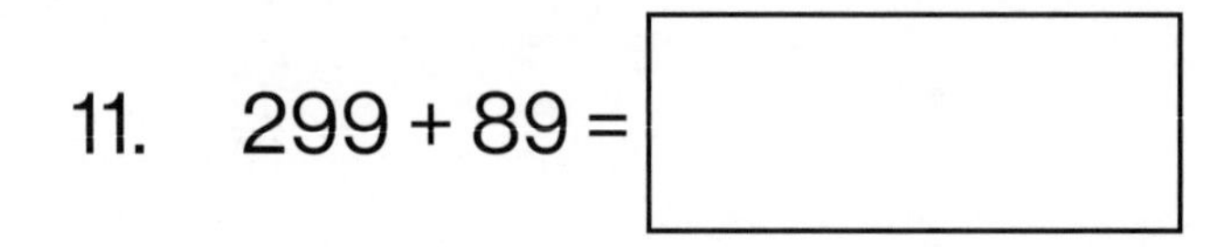

12. Find the value of 99 and 6.

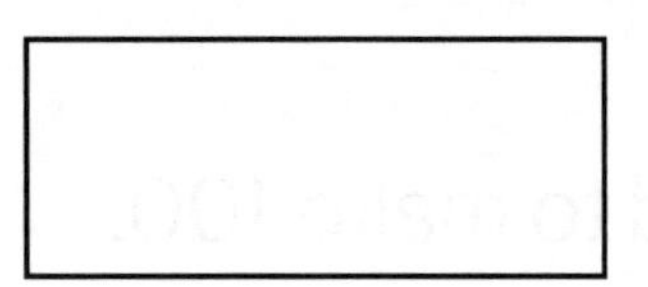

13. ☐ must be subtracted from 251 to give the answer 200.

14. ☐ tens and 10 ones make 100.

15. Write + or −.

$347 \;\square\; 25 = 322$

Name: ______________________ Date: __________

Test A

25 min

40

Score

Unit 7 Multiplication and Division

Section A (2 points each)
Circle the correct option: **A**, **B**, **C**, or **D**.

1. Multiply 7 by 4.
   What is the answer?

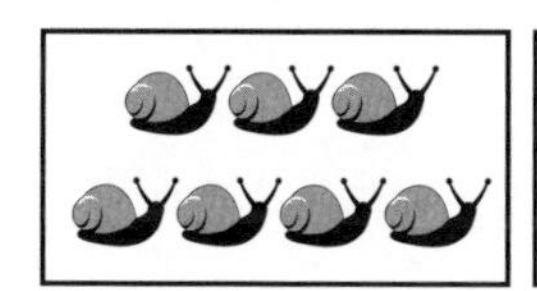 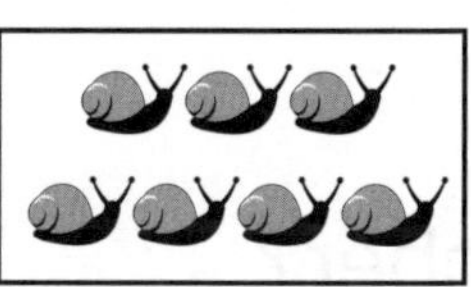 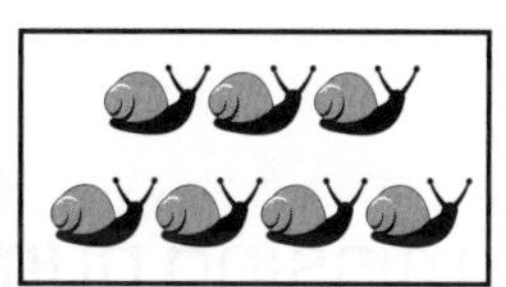 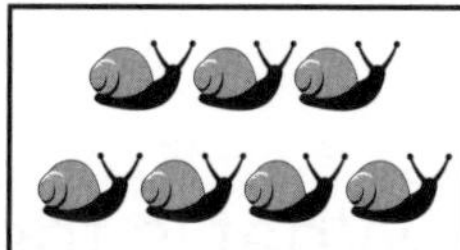

A 11 **B** 24

C 28 **D** 32

2. There are 6 sheets of stickers.
   Each sheet has 5 stickers on it.
   How many stickers are there altogether?

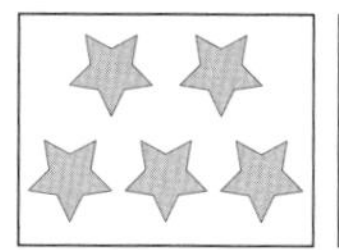 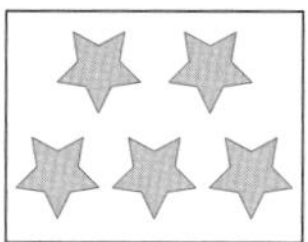 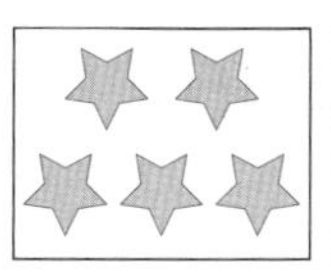 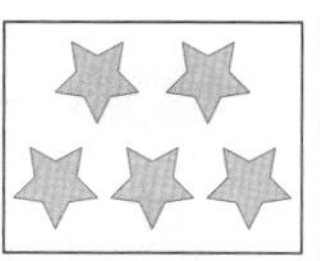 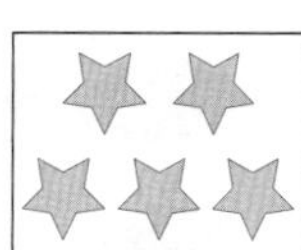 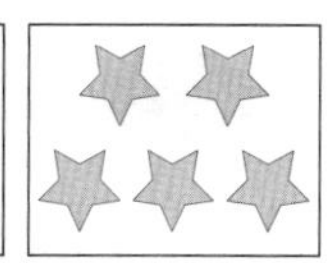

A 30 **B** 35

C 36 **D** 65

3. Kiran packed 32 apples equally into baskets.
Each basket holds 4 apples.
How many baskets of apples are there?

A 7 **B** 8

C 28 **D** 36

4. What is the missing number?

$$5 \times 4 = 4 + 4 + 4 + \boxed{?}$$

A 20 **B** 8

C 10 **D** 4

5. stands for a number.

$+ \quad + \quad = 15$

What is $\quad + \quad$?

A 3 **B** 5

C 10 **D** 15

Section B (2 points each)

6. Circle to show 5 equal groups.

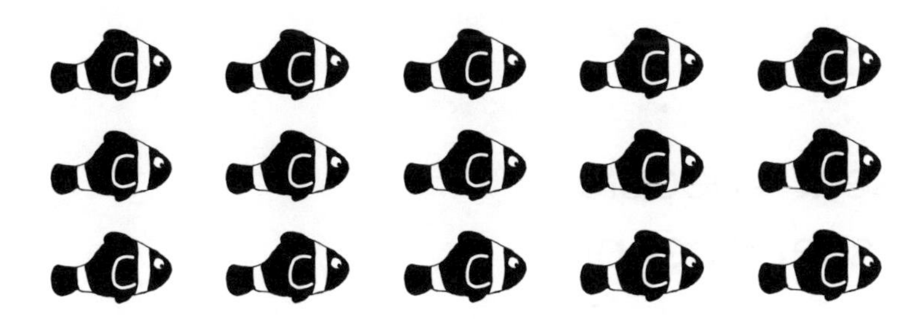

There are [ ] fish in each group.

7. Circle the box with the greatest value.

8×3	2×9	6×5	4×7

8. Look at the pattern.
Fill in the missing numbers.

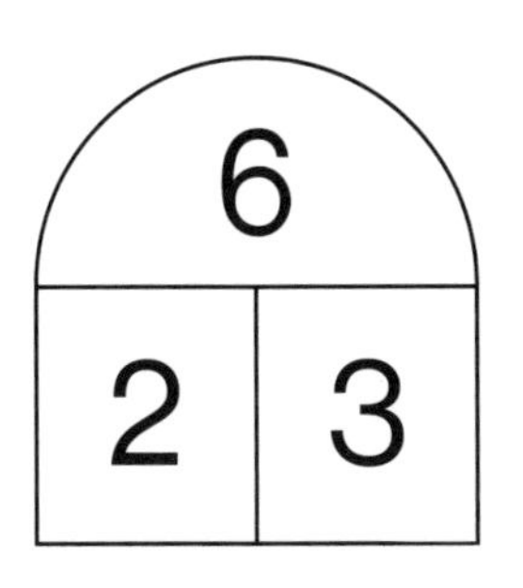

20
4 5

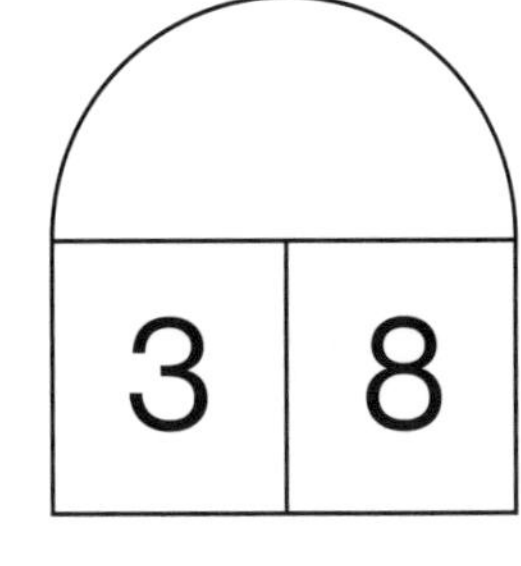

5 6

9. $5 + 5 + 5 + 5 = \square \times 5 = \square$

10. A table can seat 4 people.
2 tables can seat 8 people.

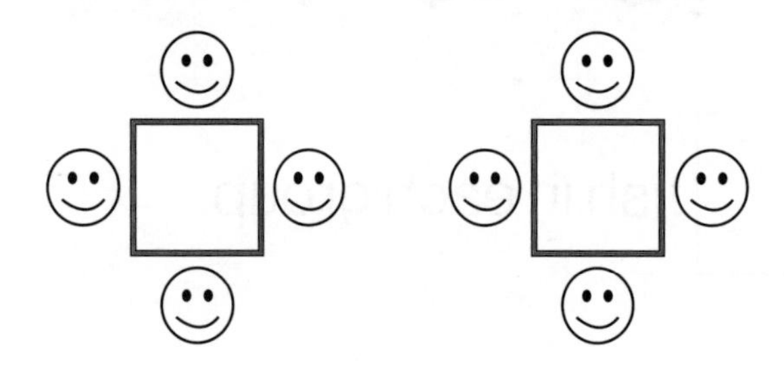

Complete the table below.

Number of tables	1	2	3		5
Number of people	4	8		16	

Section C (4 points each)

11. Each child has 5 bells.
    How many bells do 7 children have altogether?

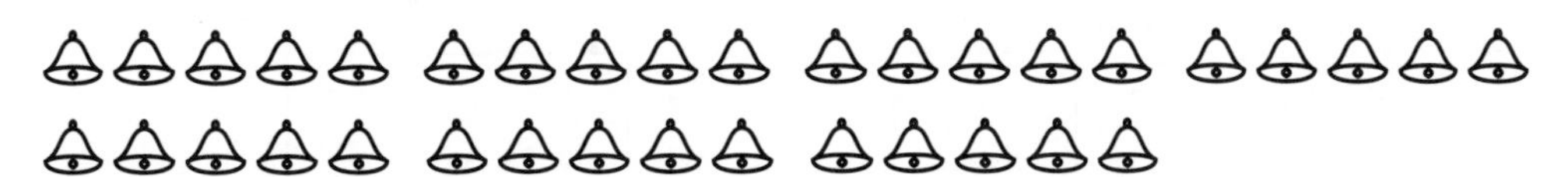

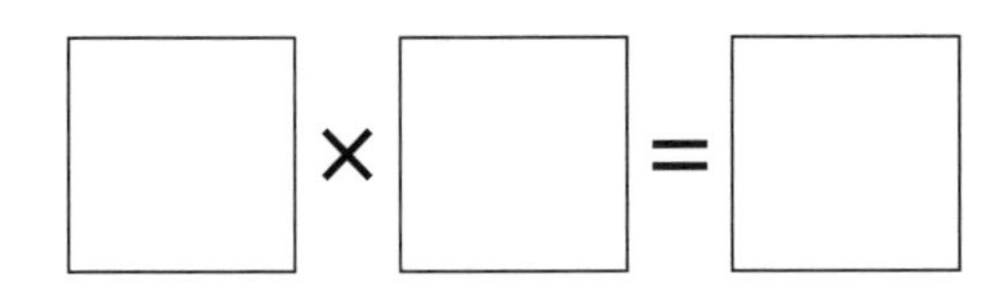

7 children have ____________ bells altogether.

12. Naomi bought 4 bags of hair clips.
    There were 5 hair clips in each bag.
    How many hair clips did she buy?

She bought ____________ hair clips.

13. Terrence bought 10 boxes of cupcakes.
Each box contained 10 cupcakes.
How many cupcakes were there altogether?

□ ○ □ ○ □

There were ____________ cupcakes altogether.

14. 30 chairs are arranged equally in 5 rows.
How many chairs are there in each row?

□ ○ □ ○ □

There are ____________ chairs in each row.

15. Andre had 6 containers of flowers.
Each container had 5 flowers.

a) How many flowers were there altogether?

There were ____________ flowers altogether.

b) Andre gave 8 flowers to his sister.
How many flowers were left?

There were ____________ flowers left.

BLANK

Name: ______________________ Date: __________

Test B

25 min

40

Score

Unit 7 Multiplication and Division

Section A (2 points each)
Circle the correct option: **A**, **B**, **C**, or **D**.

1. Manu bought 4 baskets of apples.
   There were 9 apples in each basket.
   How many apples did Manu buy?

A 11 **B** 36

C 49 **D** 50

2. Divide 20 stickers equally among 10 children.

Which equation shows the number of stickers each child receives?

A $20 + 10 = 30$ **B** $20 \div 10 = 2$

C $20 - 10 = 10$ **D** $20 \div 2 = 10$

3. What is the missing symbol?

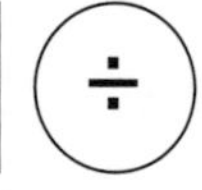 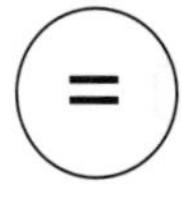 

36 ÷ 4 = 3 ? 3

A $+$ **B** $\times$

C $-$ **D** $\div$

4. Look at the number patterns.
Which set does the number '21' belong to?

A 2, 4, 6, 8, ...

B 3, 6, 9, 12, ...

C 4, 8, 12, 16, ...

D 5, 10, 15, 20, ...

5. If 🏆 + 🏆 + 🏆 = 12,
what is 🏆 × 🏆?

A 9 **B** 12

C 16 **D** 24

Section B (2 points each)

6. $6 \times 4 = 4 + 4 + 4 + 4 +$ ☐

7. 4×7 is ☐ more than 4×5.

8. Circle the box that gives the greatest answer.

$28 \div 4$	$16 \div 2$	$27 \div 3$	$30 \div 5$

9. Write the missing symbols.

15	◯	5	=	9	◯	3

10. Check (✓) the box that gives the smallest answer.

10×2	9×3	7×5	8×4
☐	☐	☐	☐

Section C (4 points each)

11. Francine bought 7 boxes of donuts.
There were 4 chocolate donuts and 6 glazed donuts in each box.

a) How many donuts were there in each box?

There were ____________ donuts in each box.

b) How many donuts did Francine buy altogether?

Francine bought ____________ donuts altogether.

12. Kyla had 36 paper clips.
She shared the paper clips equally with David, Sheena and George.

a) How many paper clips did each student get?

Each student got ____________ paper clips.

b) If David gave 5 paper clips back to Kyla, how many paper clips did Kyla have in the end?

In the end, Kyla had ____________ paper clips.

13. Fiona had 3 sticker albums.
The first two albums contained 10 stickers each.
The third album contained 7 stickers.
How many stickers did Fiona have in all?

Answer: ______________________________

14. Peter bought 36 eggs.
He used 6 eggs for baking a cake.
He then packed the remaining eggs into boxes of 5.
How many boxes of eggs were there?

Answer: ______________________________

15. Ali bought 4 bags of marbles.
There were 6 marbles in each bag.
He shared all the marbles equally with his two sisters.
How many marbles did each of them get?

Answer: ______________________________

BLANK

Name: ______________________________ Date: __________

Test A

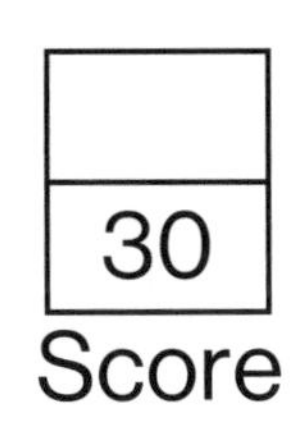

Unit 8 Money

Section A (2 points each)
Circle the correct option: **A**, **B**, **C**, or **D**.

1. What is the total amount in the box?

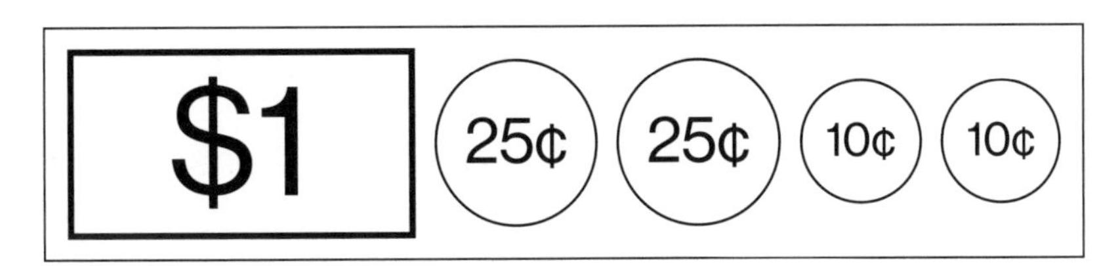

A $0.17 **B** $1.52

C $1.70 **D** $1.50

2. Elsa bought a lollipop.
She gave the cashier a $1 bill.
How much change did she get?

A 35¢ **B** 25¢

C 15¢ **D** 20¢

3. What is the cost of the calculator?

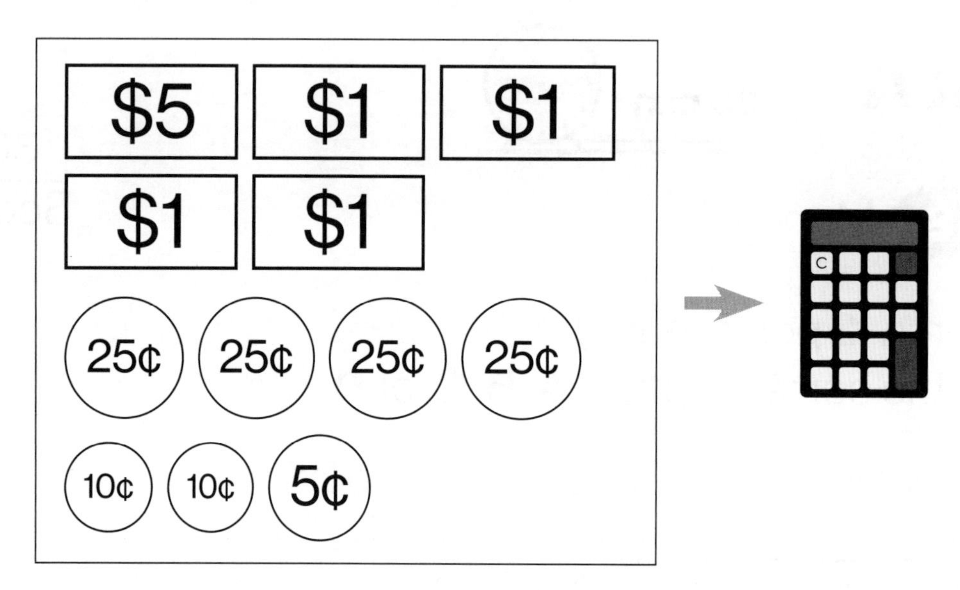

A $8.25 **B** $9.95

C $9.25 **D** $10.25

4. Which is the same as fifteen dollars and fifty cents?

A $15.15 **B** $15.50

C $50.50 **D** $50.15

5. Subtract.

$9.60 – $3.55 = ?

A $6.05 **B** $3.50

C $13.05 **D** $10.45

Section B (2 points each)

6. Write each amount of money in dollars and cents.

a) $7.05

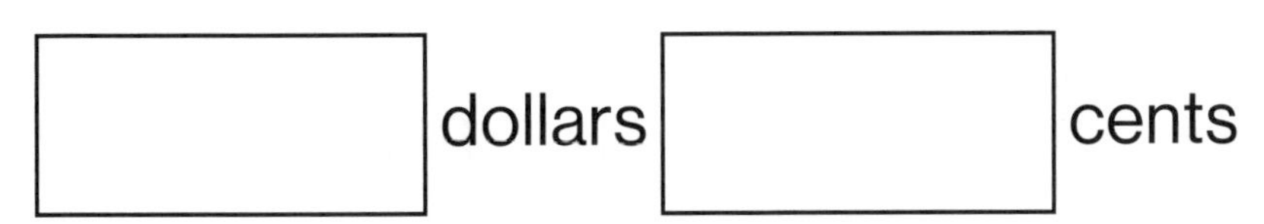

b) $13.40

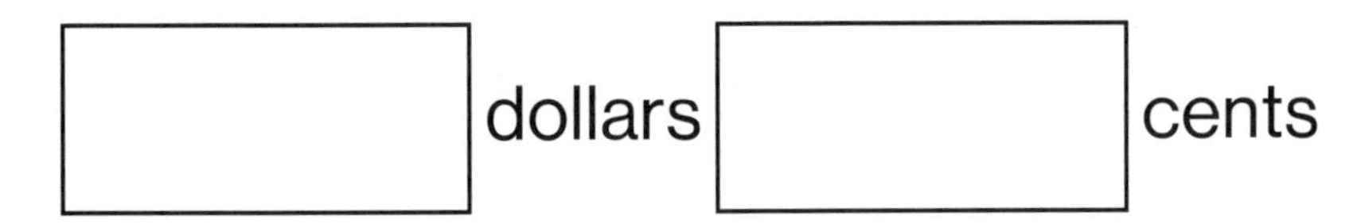

7. Write each amount of money in figures.

a) Ninety-five cents

b) Twenty-eight dollars and nine cents

$

8. The amount of money shown below is

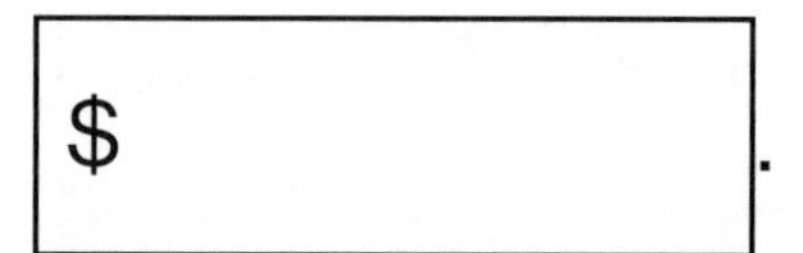

.

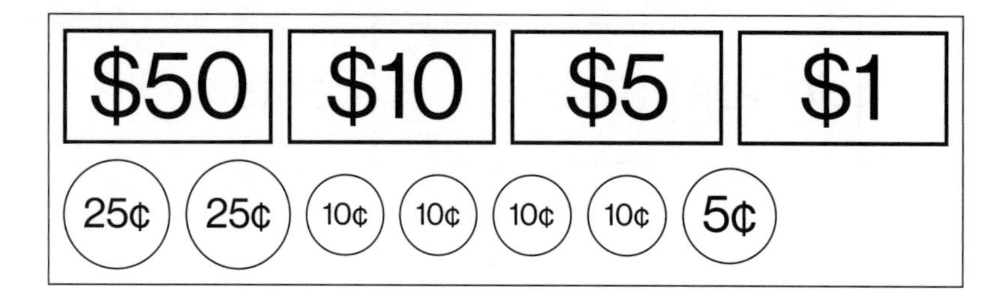

9. Color some coins below to make $1.

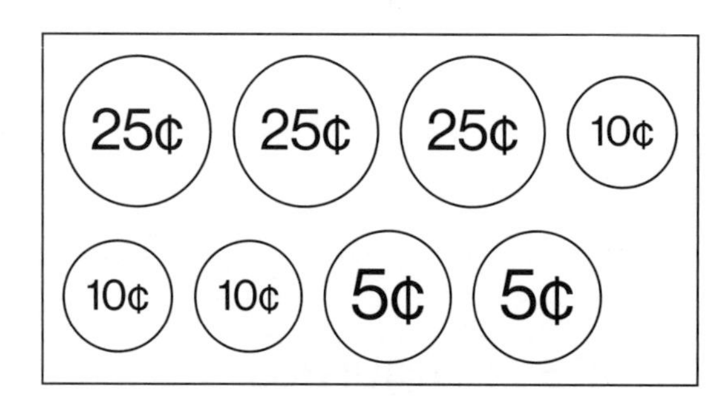

10. John spent exactly $1 on two of the fruits below. Which two fruits did he buy?

apple	kiwifruit	orange	banana
65¢	75¢	55¢	35¢

11. Add.
Fill in the boxes.

$3.75 + $6.15 = ☐

12. Subtract.
Fill in the boxes.

$8.70 – $5.40 = ☐

$8.70 —(– $5)→ ☐ —(– 40¢)→ ☐

Look at the pictures below and answer questions 13 to 15.

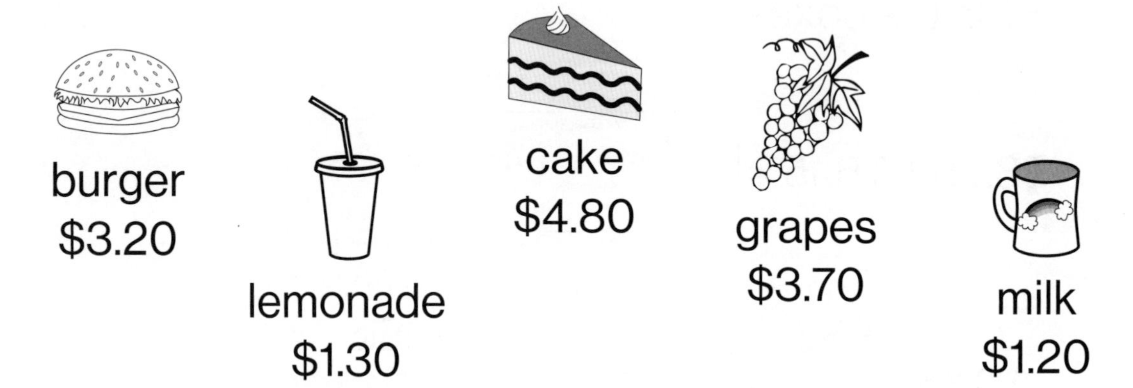

13. How much more does a burger cost than a cup of lemonade?

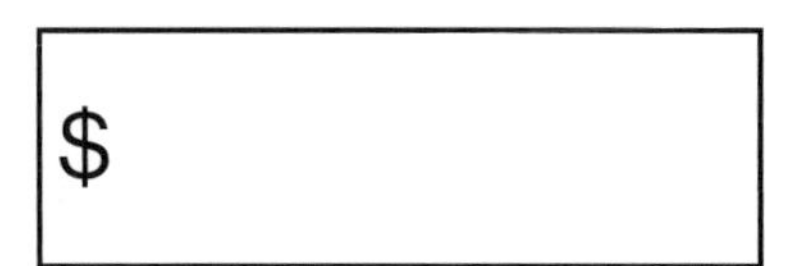

14. Suki paid $5 for two of the items.
Which two items did she buy?

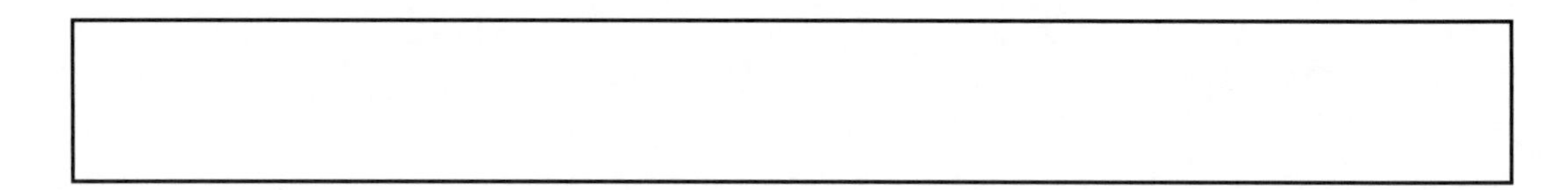

15. Bryce paid for lemonade with a $5 bill.
How much change did he get?

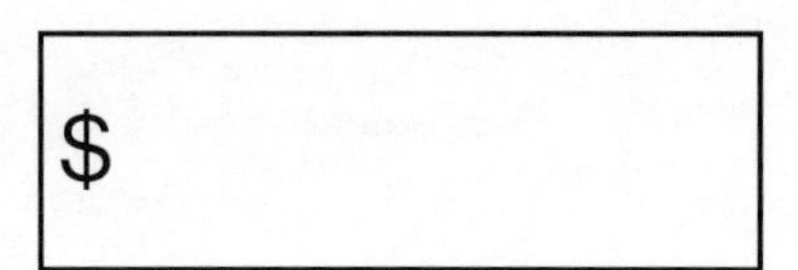

Name: ______________________________ Date: __________

Test B

30
Score

Unit 8 Money

Section A (2 points each)
Circle the correct option: **A**, **B**, **C**, or **D**.

1. How many \$5 bills are needed to make \$50?

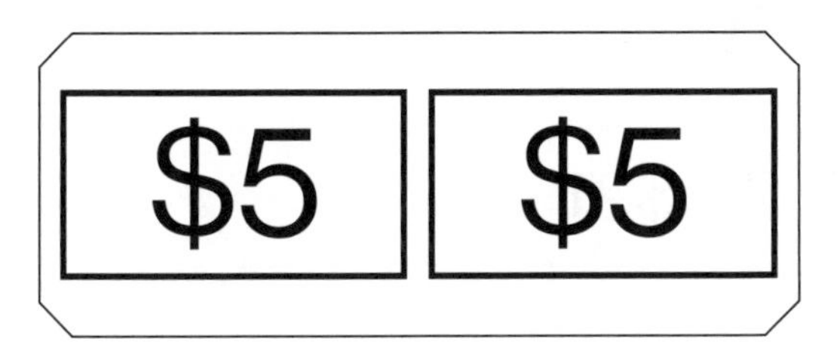

A 5 **B** 10

C 20 **D** 4

2. José has 75¢.
How much more does he need to buy these two items?

A 10¢ **B** 15¢

C 25¢ **D** 90¢

3. Monique bought a video game controller.
She gave the cashier two \$50 bills and got back \$29.
How much did the video game controller cost?

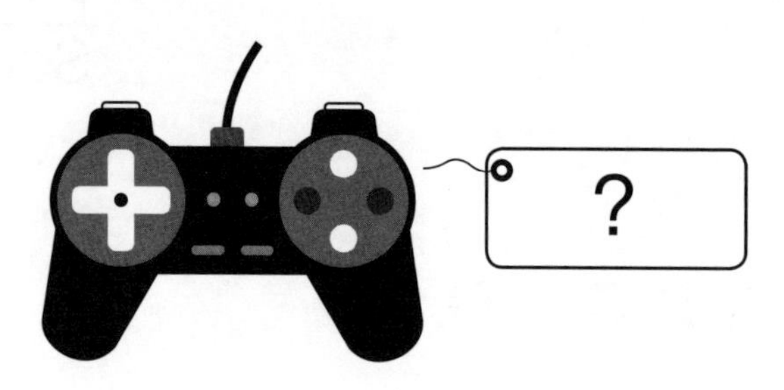

A \$29 **B** \$81

C \$71 **D** \$21

4. \$4.55 + \$3.85 = ?

A \$7.25 **B** \$7.40

C \$8.25 **D** \$8.40

5. \$6.50 − \$2.35 = ?

A \$4.15 **B** \$4.05

C \$3.85 **D** \$8.85

Section B (2 points each)

6. Count the amount of money Alyssa and James each have.
Who has less money?

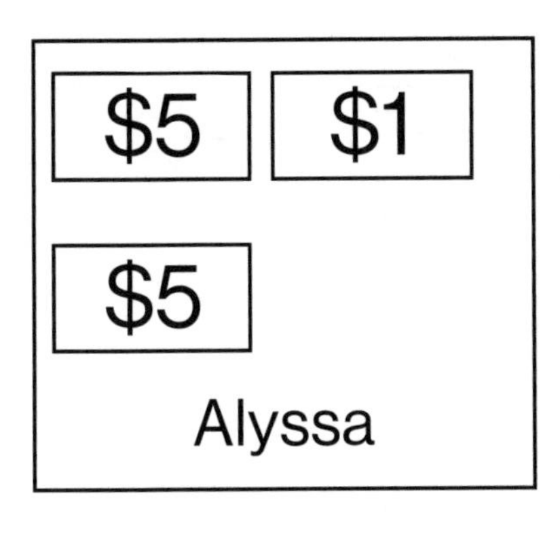

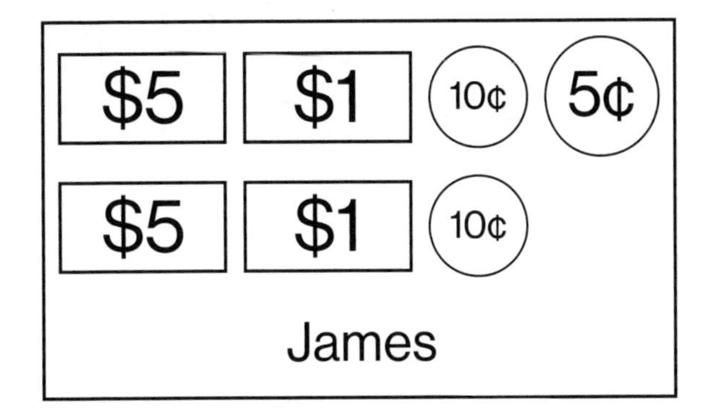

7. Agnes has 2 one-dollar bills.
She exchanges them for quarters.
How many quarters does she receive?

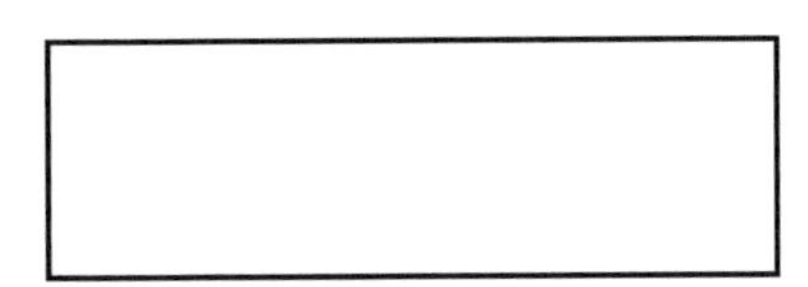

8. Color the amount of money needed to buy the robot.

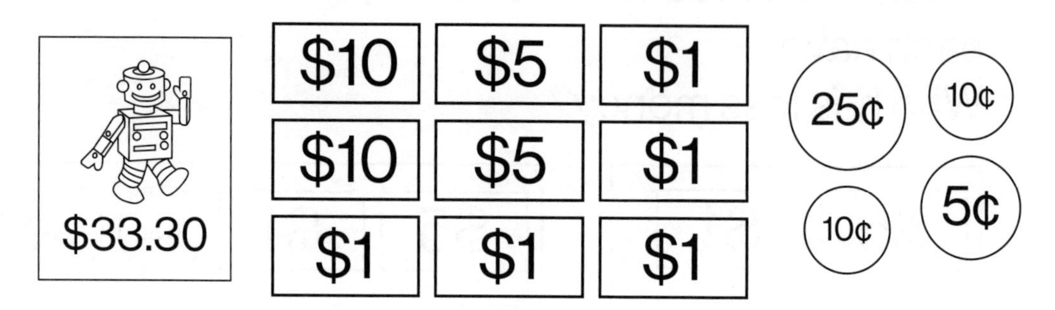

9. Naomi spent $8 on 3 of the fish below.
Circle the 3 fish that she bought.

10. Add.

$6.30 + $5.95

= $6.30 + $6 − [ ]

= $12.30 − [ ]

= [ ]

11. Subtract.

$6 − $3.95

= $6 − $4 + [ ]

= $2 + [ ]

= [ ]

Look at the pictures below to answer questions 12 to 14.

movie ticket $8.80 | soda pop $4.55 | box of candy $2.30 | popcorn $7.75

12. How much more does a movie ticket cost than a soda pop?

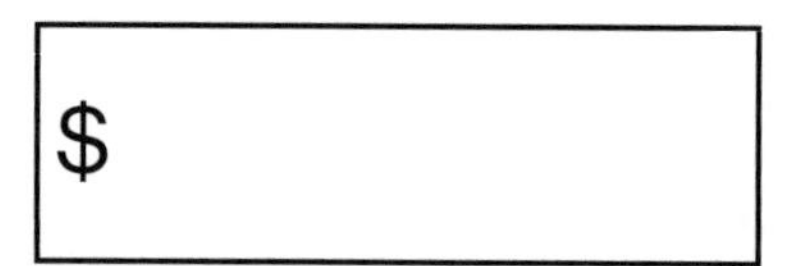

13. Abdul buys a popcorn and a box of candy.
    How much does he pay altogether?

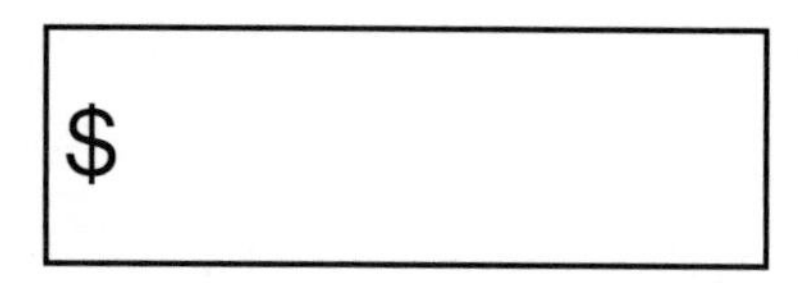

14. Violet has $5.
    How much more does she need to buy the movie ticket?

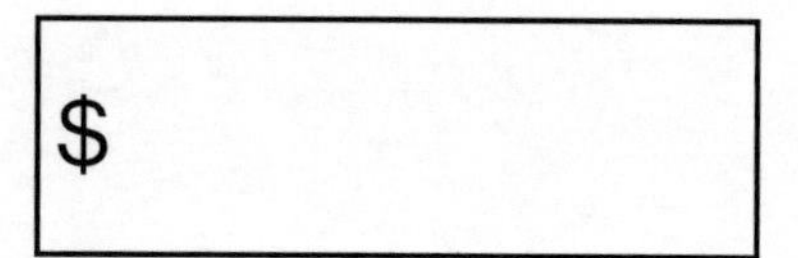

15. A watch costs $78 and a shirt costs $33.
Melvin does not have enough money and he needs $45 more to buy the two items.
How much money does Melvin have?

Answer: ______________________________

BLANK

Name: ______________________ Date: __________

Test A

25 min 

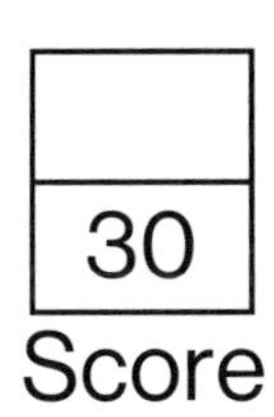

30
Score

Unit 9 Fractions

Section A (2 points each)
Circle the correct option: **A**, **B**, **C**, or **D**.

1. Which shape is divided into equal parts?

A 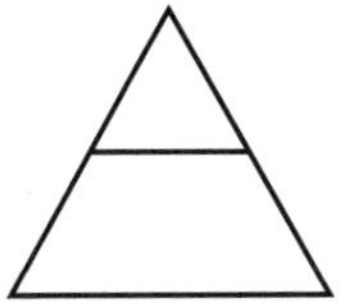

B 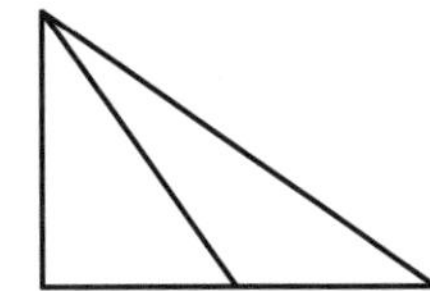

C 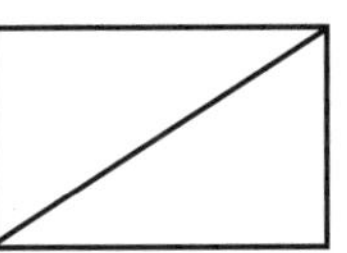

D 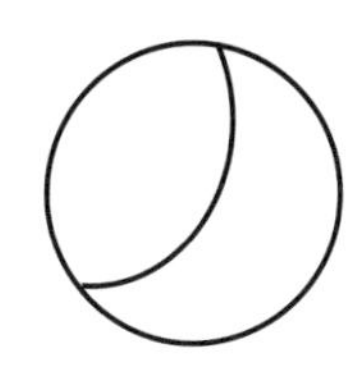

2. Which figure is $\frac{1}{4}$ shaded?

A 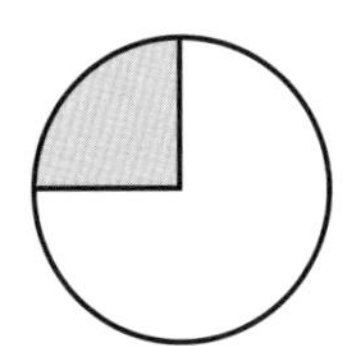

B 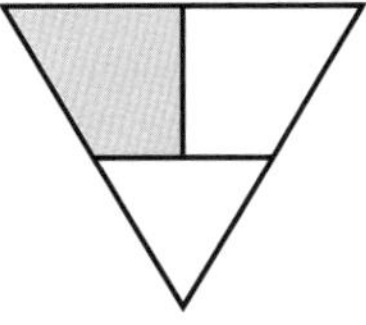

C 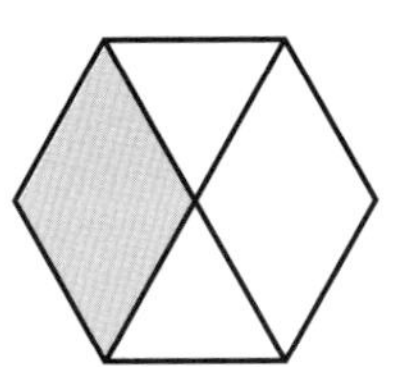

D 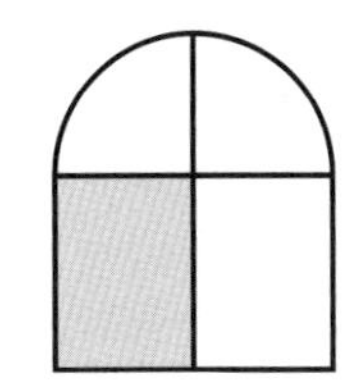

3. What fraction of the figure is shaded?

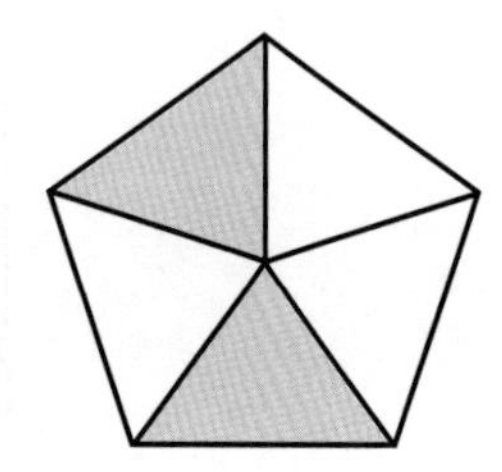

A $\frac{1}{2}$ **B** $\frac{2}{5}$ **C** $\frac{2}{3}$ **D** $\frac{3}{5}$

4. Which is the greatest fraction?

A $\frac{1}{8}$ **B** $\frac{1}{5}$ **C** $\frac{1}{3}$ **D** $\frac{1}{2}$

5. What fraction of the circle is not shaded?

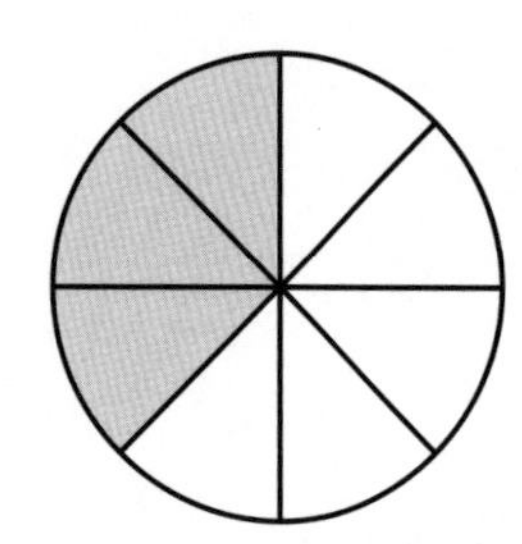

A $\frac{3}{4}$ **B** $\frac{3}{5}$ **C** $\frac{5}{8}$ **D** $\frac{3}{8}$

Section B (2 points each)

6. The figure below is divided into 4 equal parts.
What fraction of the figure is shaded?

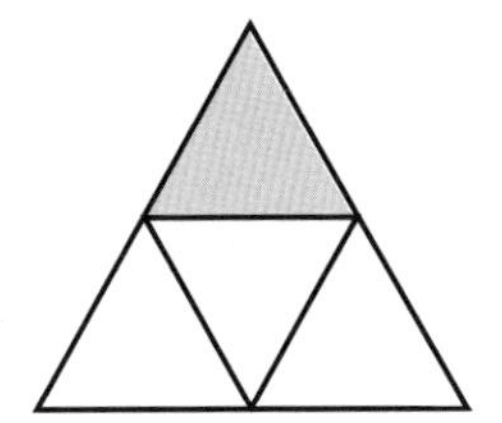

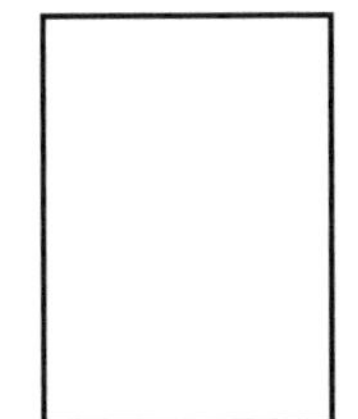

7. Arrange the fractions in order.
Begin with the greatest.

8. Arrange the fractions in order.
Begin with the smallest.

9. The figure below is divided into 6 equal parts.
   Shade $\frac{3}{6}$ of the figure.

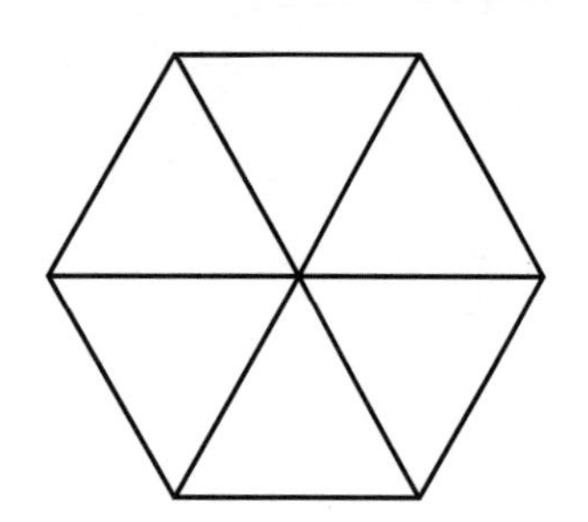

10. What fraction of the figure below is not shaded?

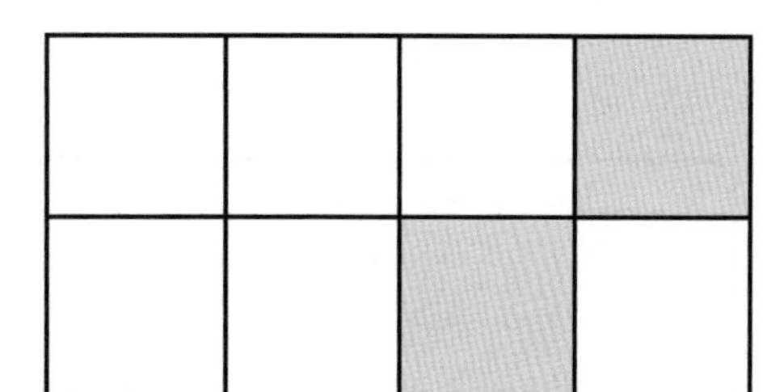

11. What fraction of the circle below is cut out?

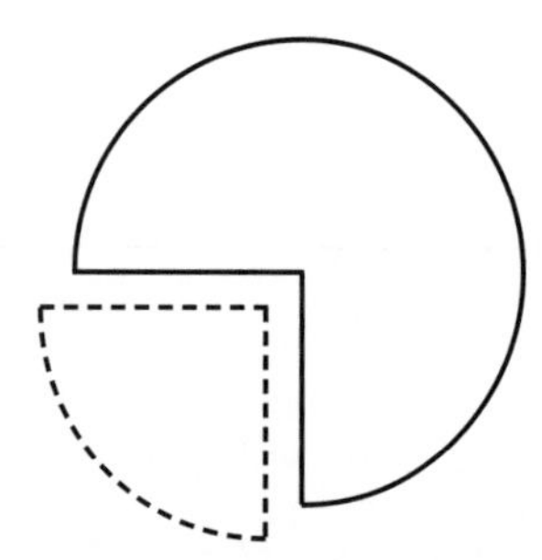

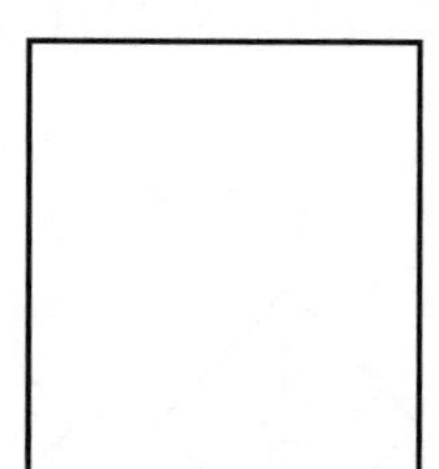

12. What fraction of the octagon below is cut out ?

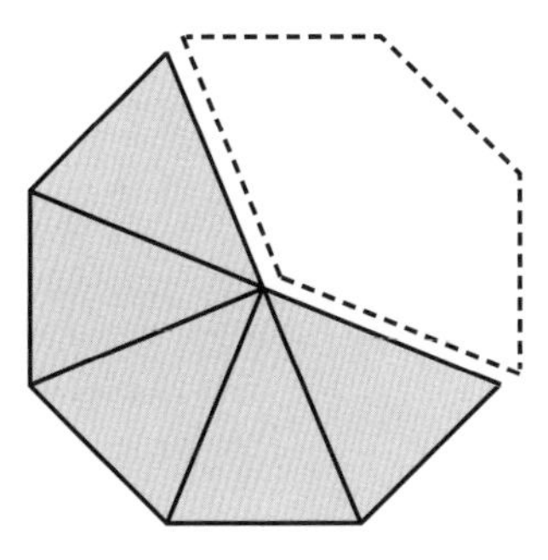

13. Diego cut a piece of wire into 5 equal parts.
He used 2 parts to make a kite.
What fraction of the wire was used?

14. $\frac{1}{5}$ and ☐ make 1 whole.

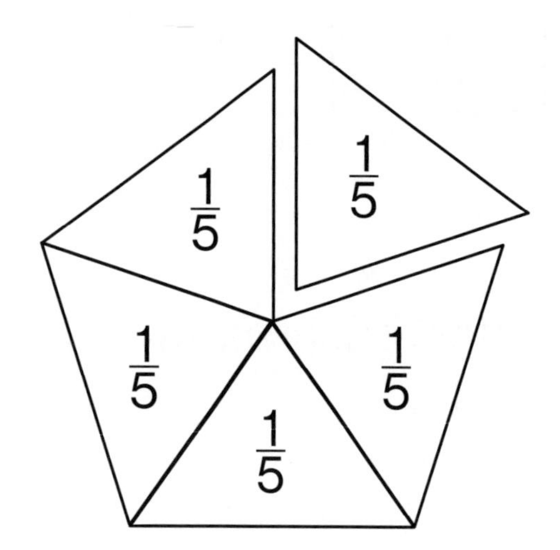

15. $\frac{2}{7}$ of a pole was painted yellow.

The rest was painted blue.

What fraction was painted blue?

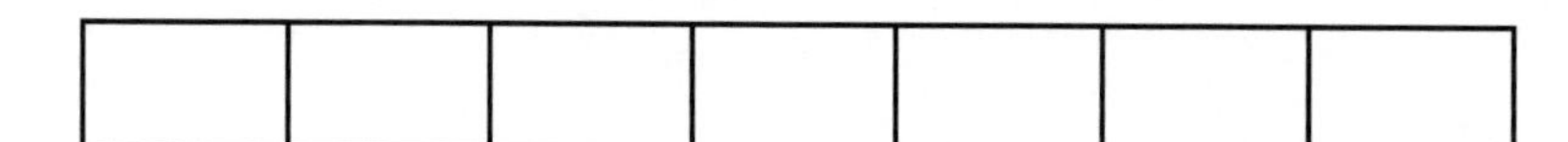

☐

Name: ______________________________ Date: __________

Test B

25 min 

30

Score

Unit 9 Fractions

Section A (2 points each)
Circle the correct option: **A**, **B**, **C**, or **D**.

1. Which figure is the odd one out?

A 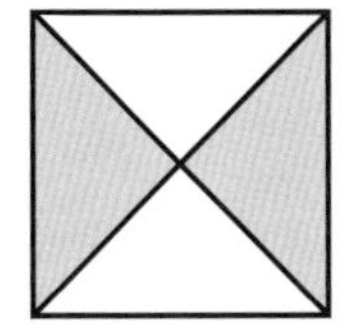

B 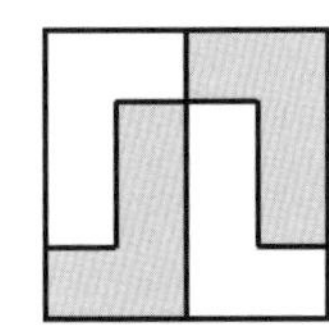

C 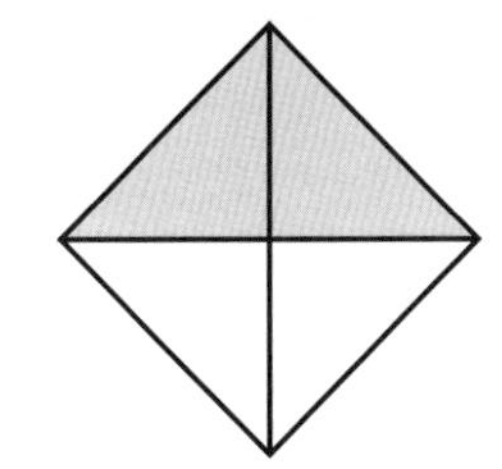

D 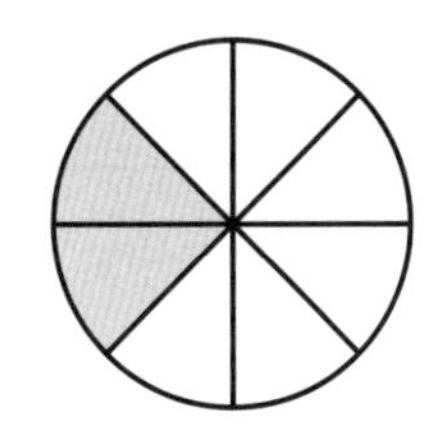

2. The figure below is made up of 6 squares.
How many squares need to be shaded to show $\frac{1}{2}$?

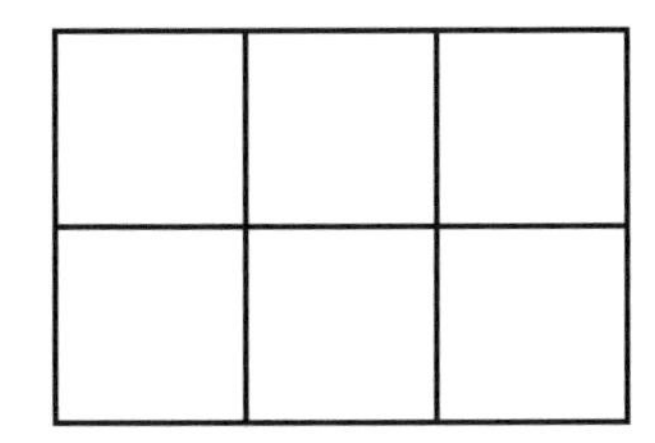

A 1

B 2

C 3

D 4

3. A is the center of the square.
What fraction of the square is shaded?

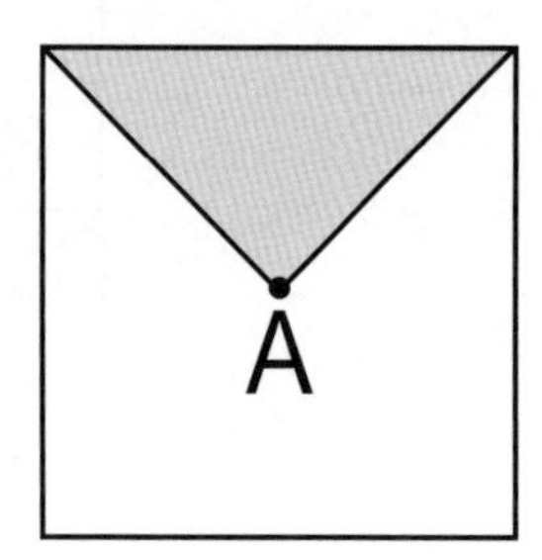

A $\frac{1}{5}$ **B** $\frac{1}{4}$ **C** $\frac{1}{3}$ **D** $\frac{1}{2}$

4. How many more squares must be shaded to show $\frac{7}{12}$?

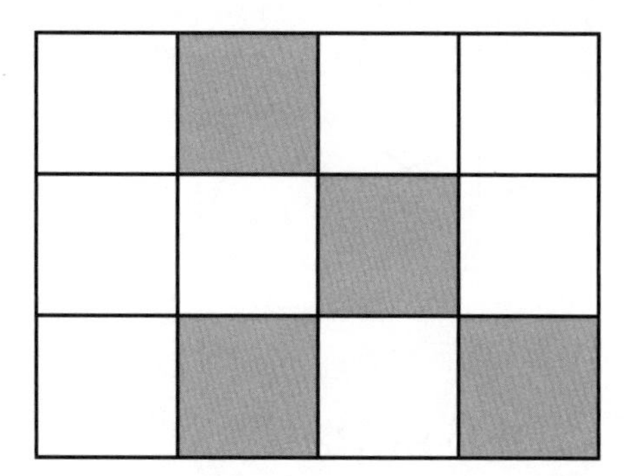

A 8 **B** 7

C 3 **D** 4

5. Tom cut a pizza into 8 equal slices.
He gave 2 slices to his brother and 1 slice to his sister.
What fraction of the pizza was left?

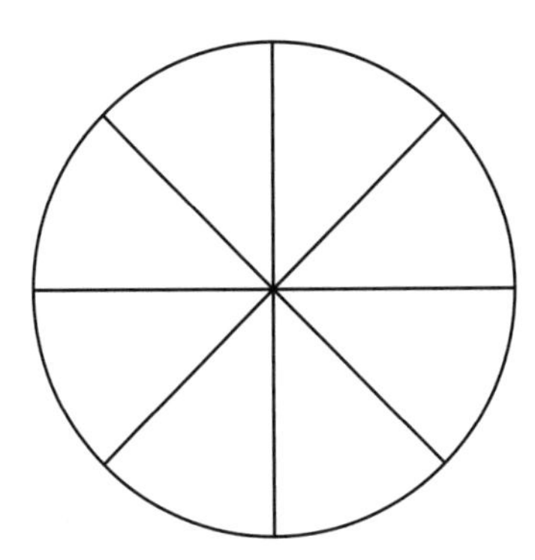

A $\frac{1}{8}$ **B** $\frac{2}{8}$ **C** $\frac{3}{8}$ **D** $\frac{5}{8}$

Section B (2 points each)

Fill in the boxes with 'smaller' or 'greater'.

6. $\frac{1}{11}$ is 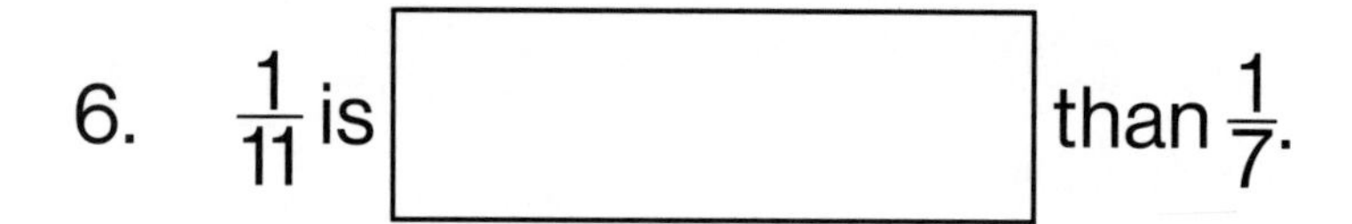 than $\frac{1}{7}$.

7. $\frac{1}{5}$ is ☐ than $\frac{1}{10}$.

8. Match each bee to a flower to make a whole.

9. Arrange these fractions in order.
Begin with the smallest.

$\frac{1}{3}$ $\frac{1}{10}$ $\frac{1}{2}$ $\frac{1}{5}$

10. How many quarters are there in 2 wholes?

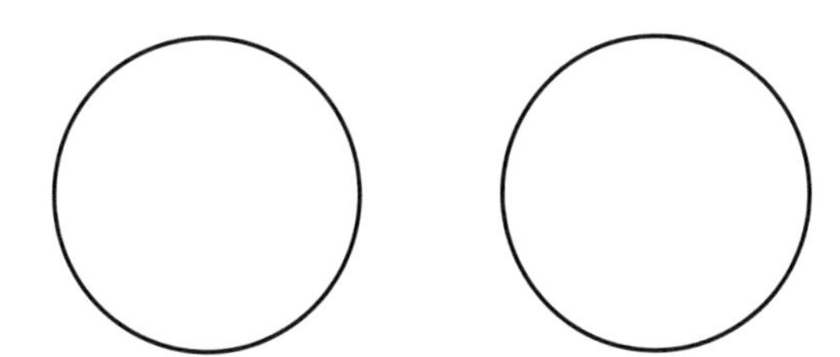

Answer: ______________________

11. What fraction of the figure is not shaded?

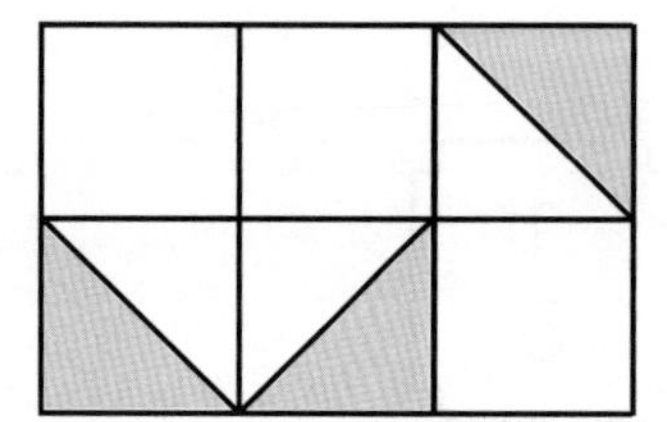

12. A chef cut a cake into 7 equal pieces.
3 customers ate 1 piece each.
What fraction of the cake was not eaten?

13. Show the fraction $\frac{5}{6}$ on the number line with an arrow (↑).

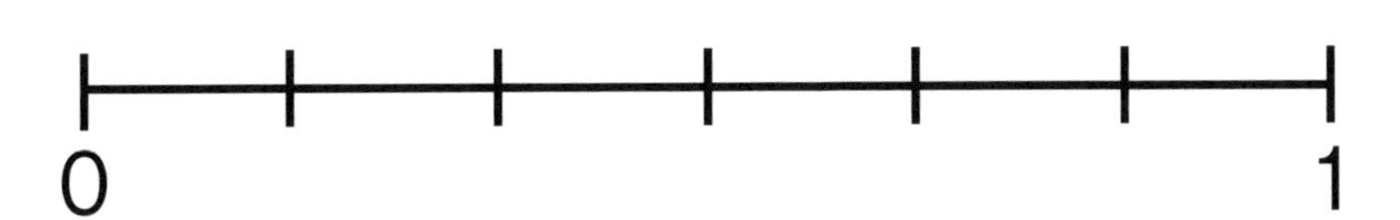

14. Ivy cut a pizza into 9 equal slices.
She ate 2 slices and gave her brother 3 slices.
What fraction of the pizza was left?

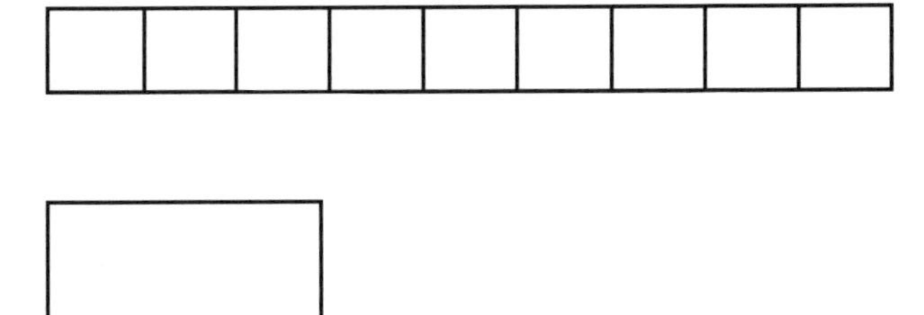

15. There are two pizzas of the same size.
Pizza A is sliced into sixths and Pizza B is sliced into eighths.
Which pizza has smaller slices?

______________ has smaller slices.

Name: ______________________________ Date: __________

Test A

45 min

60

Score

Continual Assessment 3

Section A (2 points each)
Circle the correct option: **A**, **B**, **C**, or **D**.

1. How many hundreds are in 945?

A 10 **B** 9

C 5 **D** 4

2. $\frac{3}{8}$ and __?__ make 1 whole.

A $\frac{2}{8}$ **B** $\frac{3}{8}$ **C** $\frac{5}{8}$ **D** $\frac{7}{8}$

3. Which one of the following fractions is the greatest?

A $\frac{1}{8}$ **B** $\frac{1}{3}$ **C** $\frac{1}{4}$ **D** $\frac{1}{2}$

4. Which figure shows $\frac{4}{9}$?

A 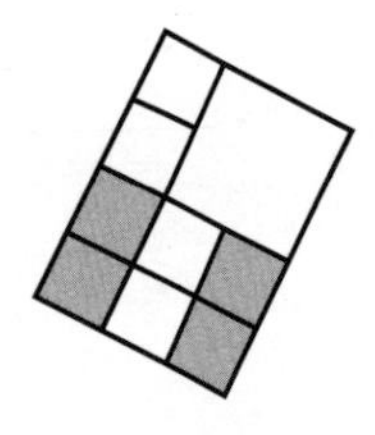

B 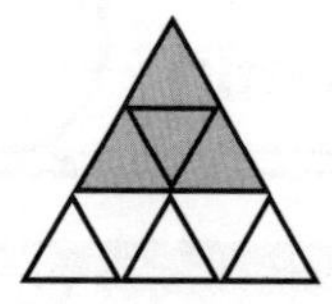

C 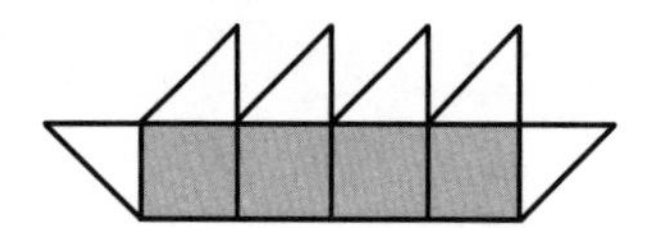

D 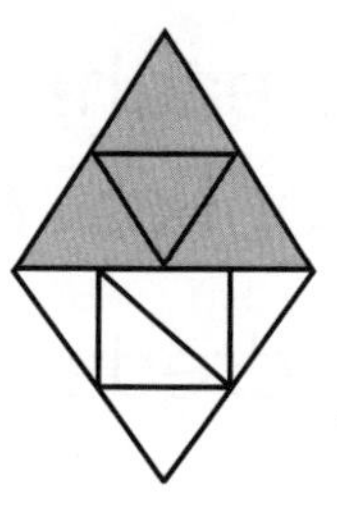

5. The missing number is ___?___.

$800 + 90 + \boxed{?} = 899$

A 1 ten

B 9 tens

C 8 tens

D 9 ones

6. What fraction of the figure is shaded?

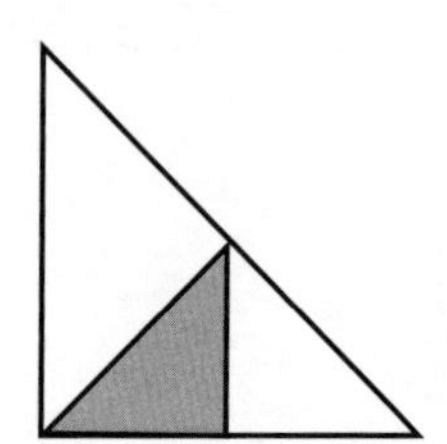

A $\frac{1}{4}$ **B** $\frac{1}{3}$ **C** $\frac{2}{3}$ **D** $\frac{3}{4}$

7. The length of the pencil is __?__ cm.

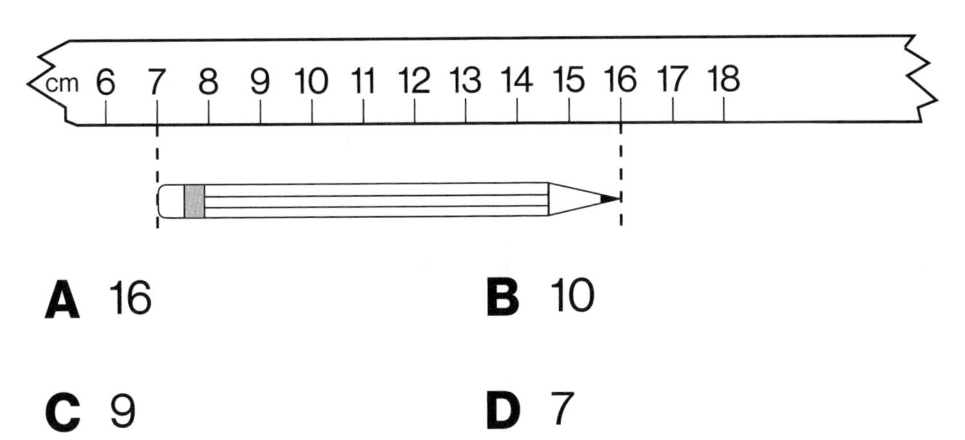

A 16 **B** 10

C 9 **D** 7

8. A chocolate bar was divided into 7 equal pieces.
Lynn ate 1 piece of the chocolate bar.
What fraction of the chocolate bar did she eat?

A $\frac{1}{7}$ **B** $\frac{2}{7}$ **C** $\frac{5}{7}$ **D** $\frac{6}{7}$

9. Adam bought a muffin and a bottle of juice for breakfast.
The muffin cost $1.75 and the juice cost $2.55.
What was the total cost of Adam's breakfast?

A $5.30 **B** $3.85

C $4.40 **D** $4.30

10. Julie and her brother saved money for a month.
Julie saved $9.60 and her brother saved $5.10.
How much more money did Julie save than her brother?

A $3.50 **B** $5.40

C $4.50 **D** $6.50

Section B (2 points each)

11. The value of the digit 8 in 853 is ☐.

12. Write in words.

679 ☐

13. Complete the number pattern below.

14. Use the given digits to form the smallest 3-digit odd number.

7 0 3 → ☐

15. \$7.20 = ☐ cents

16. Thirty-seven cents = \$ ☐

17. Put 18 marbles equally into 3 groups.

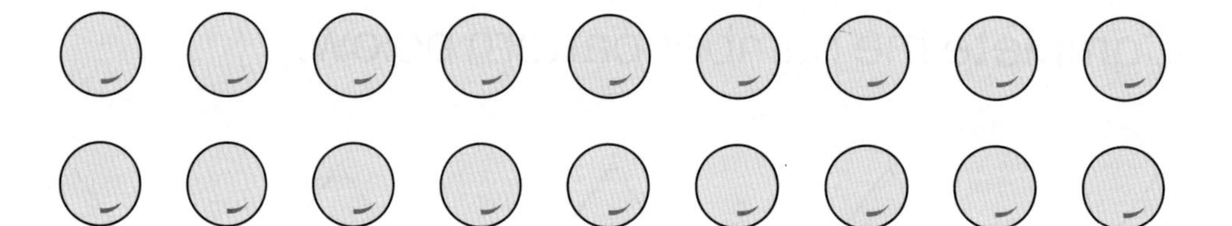

There are ____________ marbles in each group.

18. 6 groups of 10 is the same as 10 × ☐.

19. Each student has 8 storybooks.

3 students have ____________ storybooks altogether.

20. Grace has a piece of lace 18 m long.
She needs to sew 3 m of lace on each dress.
How many dresses can she sew lace on?

Grace can sew lace on ____________ dresses.

Section C (4 points each)

21.

orange 50¢ each 	apple 75¢ each 

a) An orange costs ____________ ¢ less than an apple.

b) The total cost of the two fruits is

$ ____________ .

22. Ashton cut a piece of paper into 8 equal pieces.
He used 5 pieces.

a) What fraction of the paper did Ashton use?

Ashton used ____________ of the paper.

b) What fraction of the paper was left?

There was ____________ of the paper left.

23. Lily walks towards a bus stop 260 m away.
She stops for a rest after walking 115 m.
How far is she from the bus stop?

She is ____________ from the bus stop.

24. Jenna spent $8.30 and Marcus spent $6.55.
How much more did Jenna spend than Marcus?

Jenna spent ____________ more than Marcus.

25. Rose arranges 6 paper clips in a line.

The line of paper clips has a length of

______________ cm.

Extra Credit

1. Study the patterns.

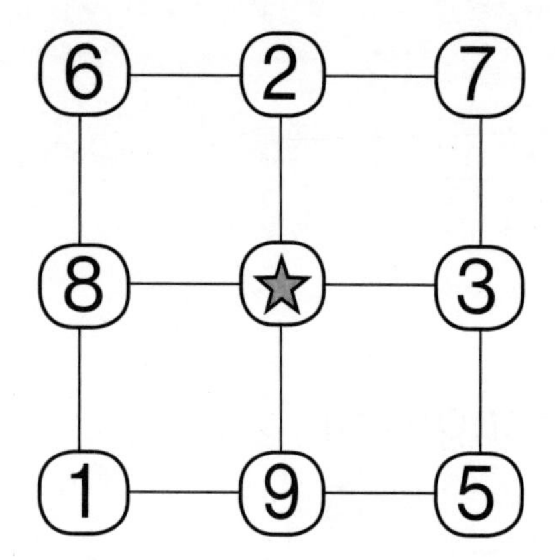

★ stands for ____________.

2. In a classroom, 6 chairs are placed between every 2 tables.
   If there are 3 tables, how many chairs are there altogether?

There are ____________ chairs altogether.

Name: ______________________________ Date: __________

Test B

45 min

60

Score

Continual Assessment 3

Section A (2 points each)
Circle the correct option: **A**, **B**, **C**, or **D**.

1. 4×5 is the same as __?__.

   A $4 + 4 + 4 + 4$ **B** $5 + 5 + 5 + 5$

   C $5 \times 5 \times 5 \times 5$ **D** $4 \times 4 \times 4 \times 4$

2. What is the missing sign?

   267 [?] 117 = 150

   A $+$ **B** $-$

   C $\times$ **D** $\div$

3. What is the missing number?

   [?] − 421 = 79

   A 342 **B** 490

   C 500 **D** 510

4. How many tens are there in 5×8?

A 1 **B** 10

C 40 **D** 4

5. $\frac{2}{7}$ and ___?___ make 1 whole.

A $\frac{2}{7}$ **B** $\frac{3}{7}$ **C** $\frac{4}{7}$ **D** $\frac{5}{7}$

6. $\frac{2}{10}$ is 2 out of ___?___ equal parts.

A 1 **B** 2 **C** 12 **D** 10

7. Which figure shows $\frac{1}{2}$?

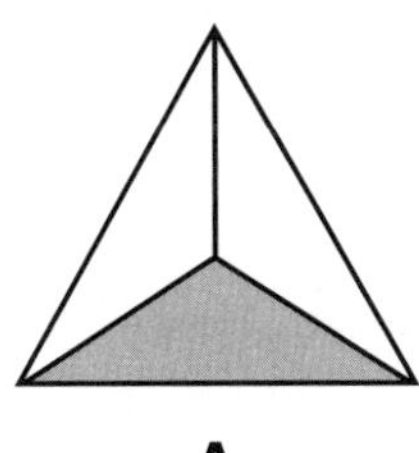

A

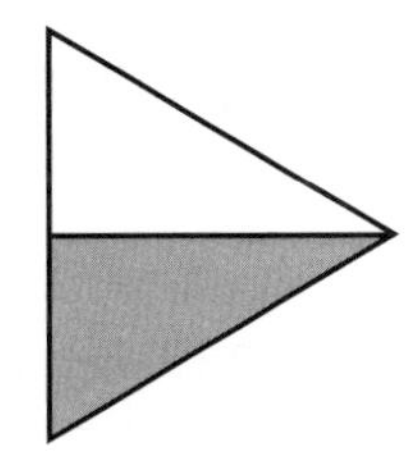

B

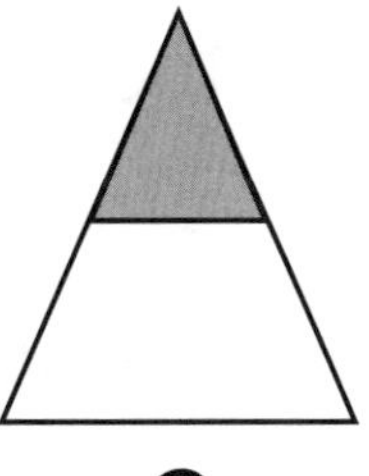

C

A A

B B

C C

D None of them

8. What fraction of the figure is shaded?

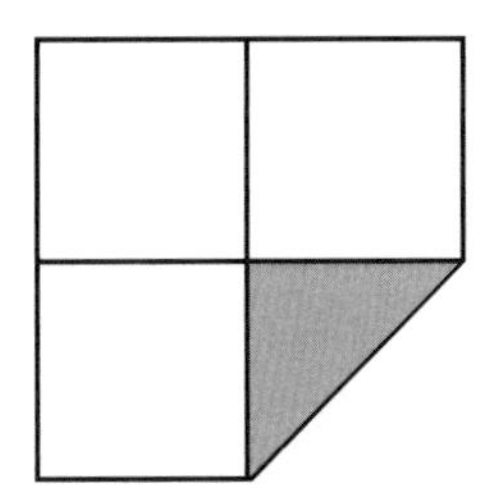

A $\frac{1}{2}$ **B** $\frac{1}{3}$ **C** $\frac{1}{7}$ **D** $\frac{1}{8}$

9. Which figure is $\frac{2}{3}$ shaded?

A 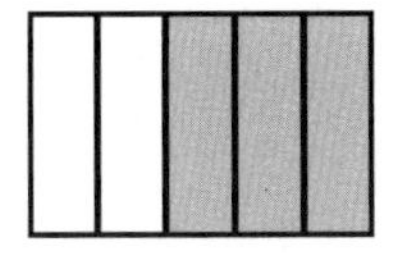

B 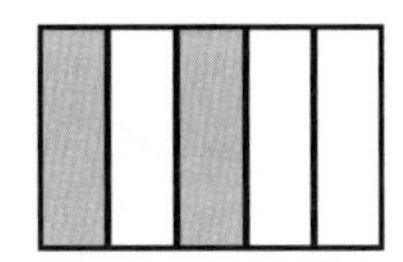

C 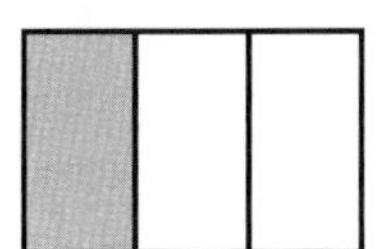

D 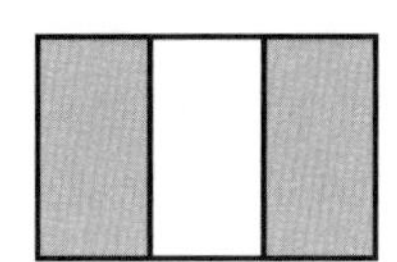

10. The total length of the greeting cards below is 40 cm.
What is the length of each greeting card?

A 8 cm **B** 5 cm **C** 4 cm **D** 10 cm

Section B (2 points each)

11. Arrange these numbers in order.
Begin with the greatest.

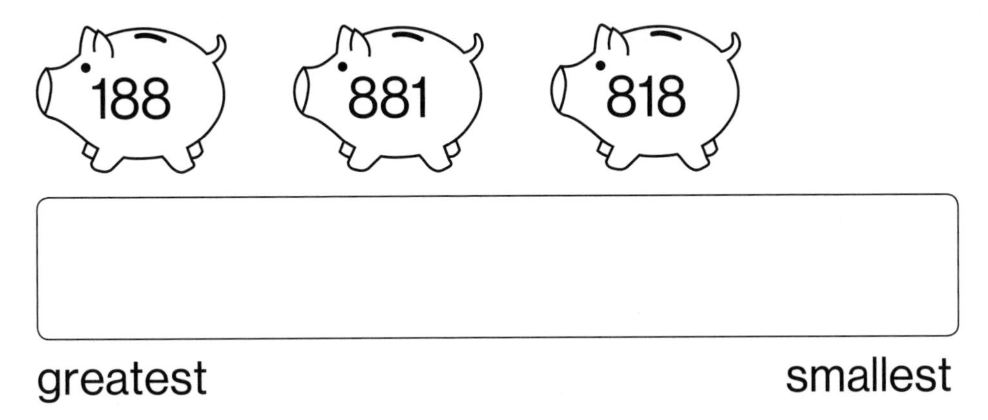

12. 325 + ? = 900

The missing number is ______.

13. 9 + 9 + 9 + 9 = ? × 4

The missing number is ______.

14. $30 \div 3 = 2$ [?] 5.

The missing sign is [ ].

15. Arrange the fractions in order.
Begin with the smallest.

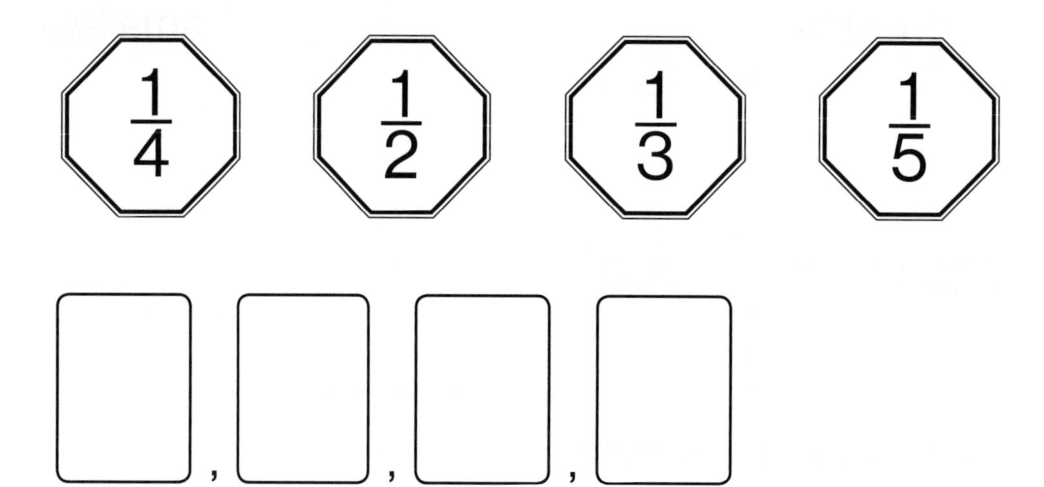

16. [ ] of the figure is not shaded.

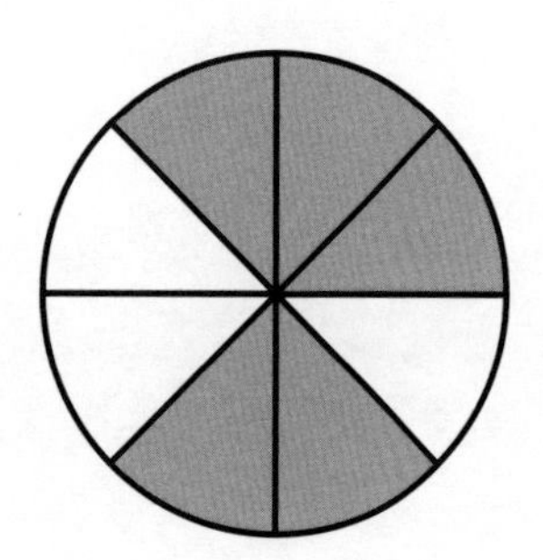

17. Use the 3 digits to write the least amount of money.

6, 0, 2 → $ ______

18. Look at the picture below.

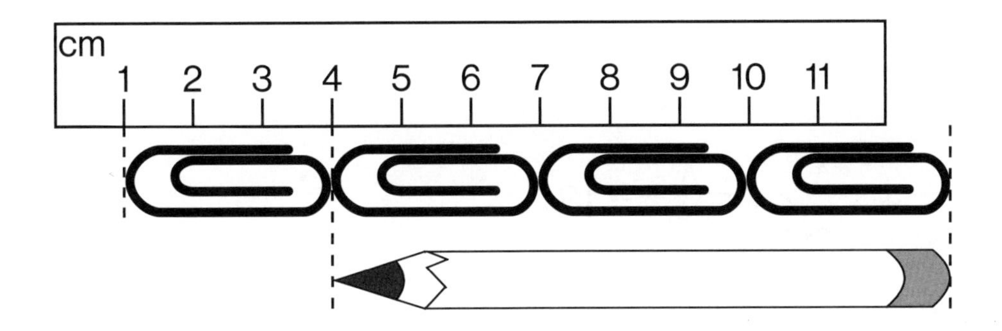

a) Each paper clip is ______________ cm long.

b) The pencil is ______________ cm long.

19. A deer has 4 legs.

7 deer have ____________ legs.

20. ★ + ★ + ★ = 18

♥ + ★ = 10

♥ = [ ]

Section C (4 points each)

21. A sub sandwich was cut into 7 equal slices.
    Johan and Rita ate 2 slices each.
    What fraction of the sandwich was left?

    Answer: ____________________

22. There are 2 red beads and 3 black beads in each jar.
    How many beads are there in 4 jars?

    There are __________ beads in 4 jars.

23. A folder costs $1.25.
A scrapbook costs $2.50 more than the folder.

a) How much does the scrapbook cost?

The scrapbook costs ____________.

b) How much do the folder and the scrapbook cost altogether?

The folder and the scrapbook cost

____________ altogether.

24. Lauren used 29 yd of ribbon to tie presents. She used 3 yd of ribbon for each present.

a) How many presents did she tie?

She tied ___________ presents.

b) How much ribbon did she have left over?

She had ___________ left over.

25. Amy wants to buy 4 burgers.
She has $10 in her purse.
How much more money does she need?

$

Extra Credit

1.

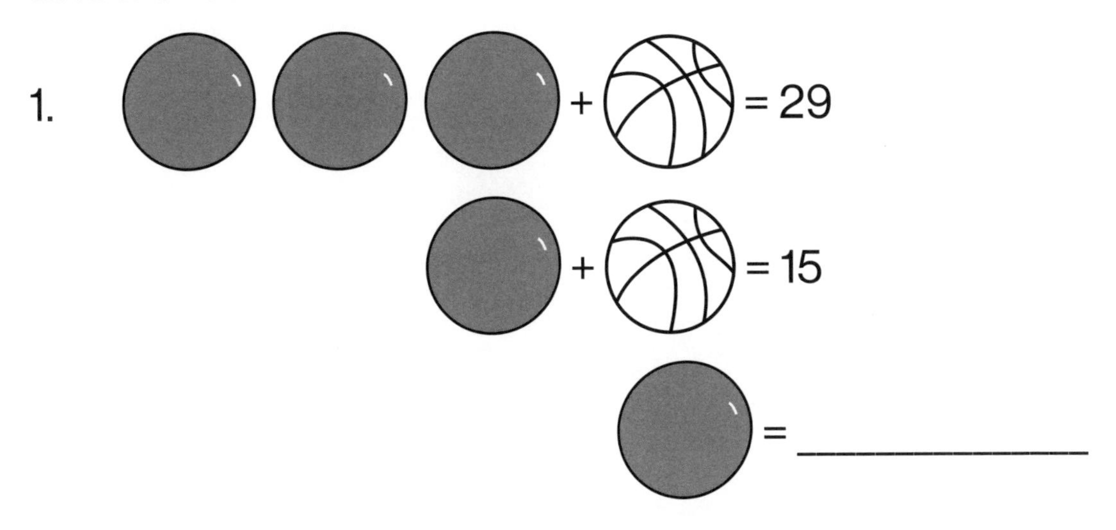

2. A doll costs $6.
2 dolls cost the same as 3 kites.
What is the cost of each kite?

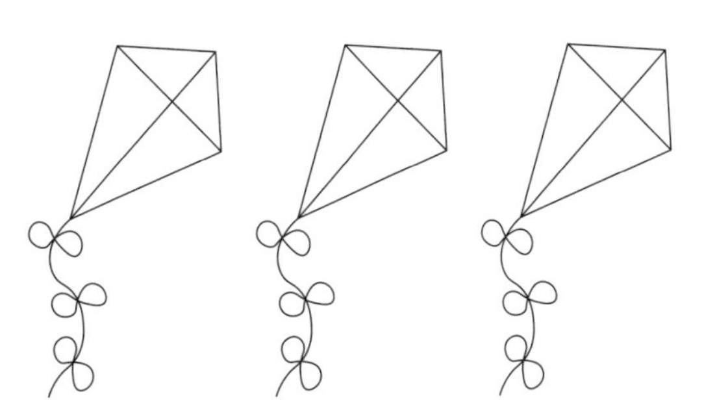

The cost of each kite is ______________.

BLANK

Name: ______________________ Date: __________

Test A

30

Score

Unit 10 Time

Section A (2 points each)
Circle the correct option: **A**, **B**, **C**, or **D**.

1. What is the time shown on the clock?

   A 8:30

   B 8:15

   C 4:40

   D 3:40

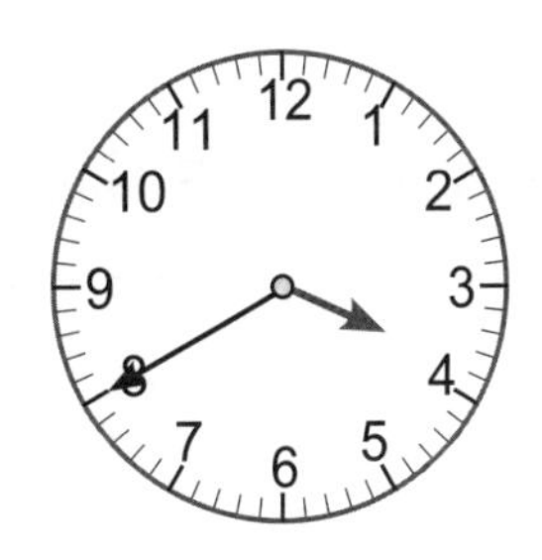

2. What time is shown on the clock?

   A 20 minutes before 7 o'clock

   B 20 minutes before 8 o'clock

   C 20 minutes after 7 o'clock

   D 20 minutes after 8 o'clock

3. What is the time shown on the clock?

A 40 min before 10 o'clock

B 10 min before 8 o'clock

C 20 min before 10 o'clock

D 50 min before 8 o'clock

4. The clock shows the time Raja wakes up in the morning. What time does Raja wake up each morning?

A 20 minutes to 6

B 20 minutes past 6

C 20 minutes to 7

D 20 minutes past 7

5. Which clock will show 5 o'clock soon?

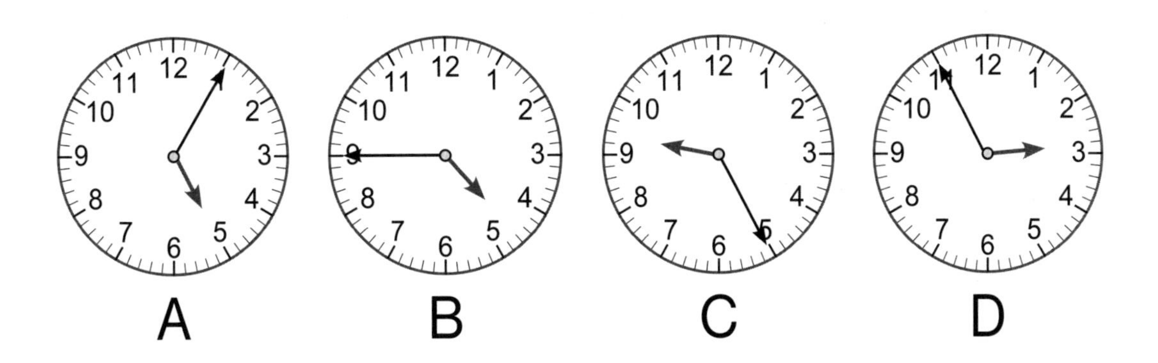

A A

B B

C C

D D

Section B (2 points each)

6. What time is shown on the clock?

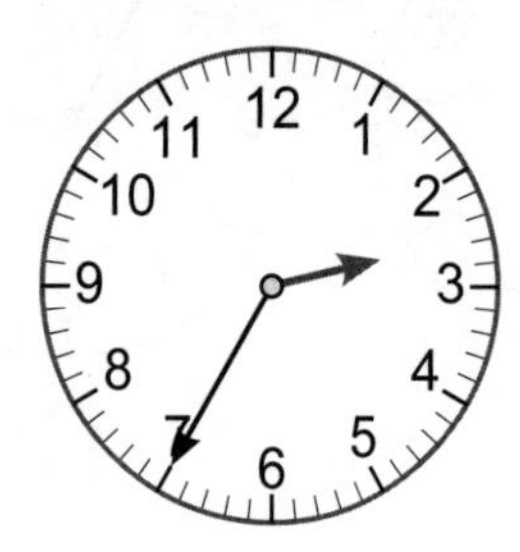

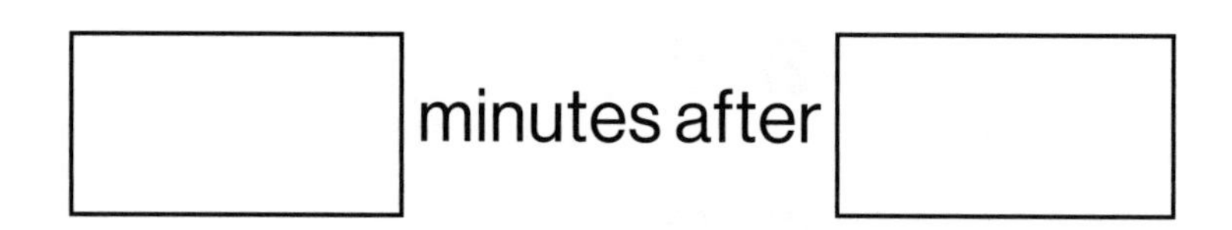

7. The time shown on the clock is ______________.

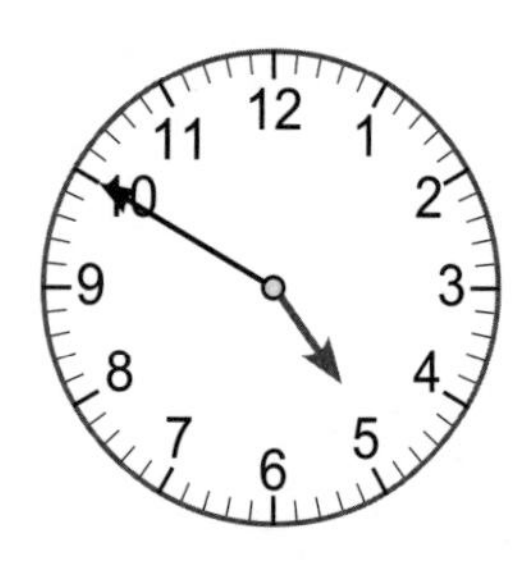

8. What time is shown on the clock?

☐ minutes after ☐

9. The clock shows the time Jolie gets up in the morning.
At what time does Jolie get up?

Answer: ______________________

10. Draw the missing minute hand to show 45 minutes past 4.

11. Draw the missing minute hand to show 15 minutes to 3.

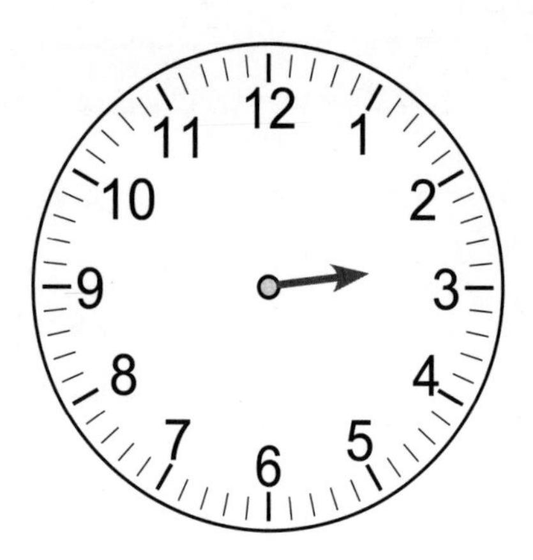

12. Draw the missing minute hand to show 9:10 A.M.

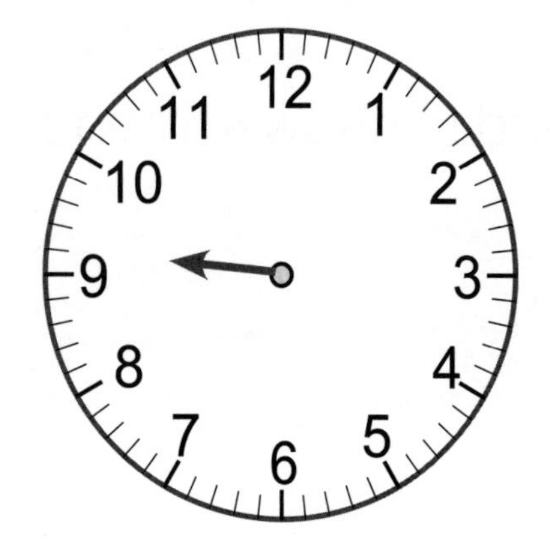

13. Draw the missing hour hand to show 12:30.

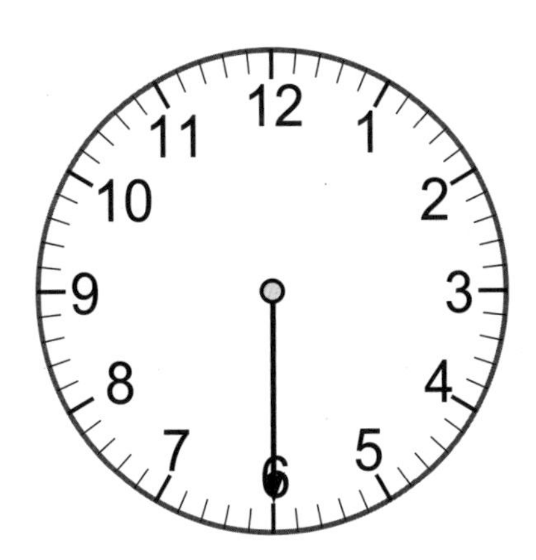

14. Mark starts reading every night at the time shown below.
At what time does he start reading?
(Use A.M. or P.M.)

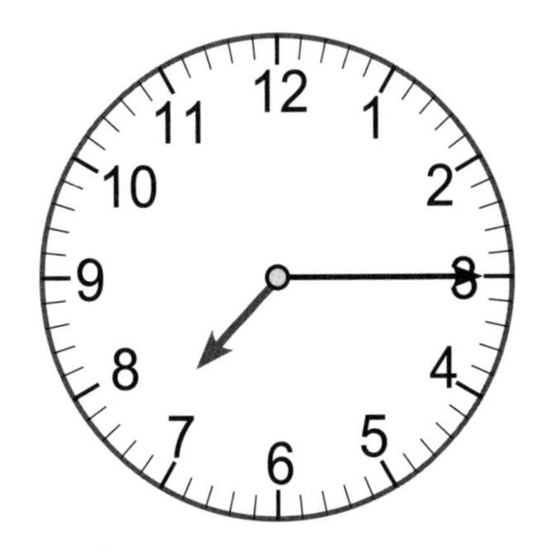

15. Cindy wakes up every morning at the time shown below.
At what time does she wake up (use A.M. or P.M.)?

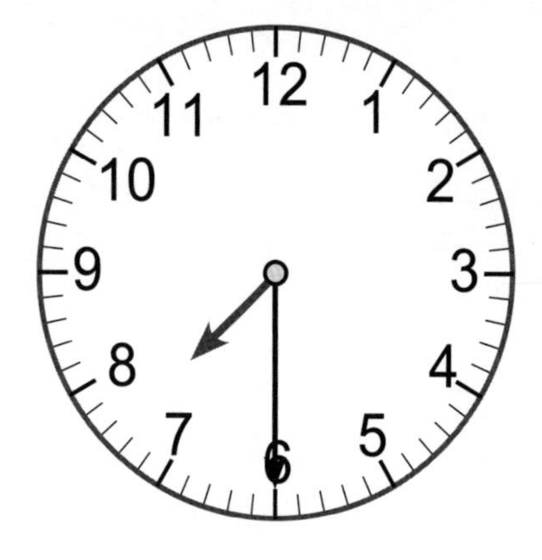

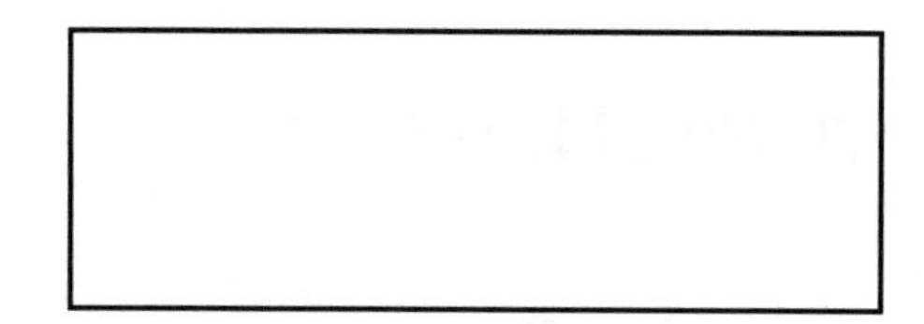

Name: ______________________________ Date: __________

Test B

25 min

30

Score

Unit 10 Time

Section A (2 points each)
Circle the correct option: **A**, **B**, **C**, or **D**.

1. What is the time shown on the clock?

A 3:00 A.M.

B 3:15 P.M.

C 15 minutes after 2 o'clock

D 3 o'clock

2. What time did Kenya wake up this morning?

A 1:00 P.M.

B 7:00 A.M.

C 7:00 P.M.

D noon

3. Ten minutes before noon is the same as what time?

A 12:10 P.M.

B 11:50 P.M.

C 11:50 A.M.

D 12:10 A.M.

4. Pat took his dog for a walk after lunch.
At what time did he go for his walk?

A 12:00 P.M.

B 7:00 A.M.

C 7:30 A.M.

D 6:30 A.M.

5. Which clock shows 20 minutes past 9?

A

B

C

D

Section B (2 points each)

6. What time is shown on the clock?

The time is ____________ minutes past 5.

7. Write the time shown on the clock.

____________ minutes past 6.

8. Fill in the blank with **A.M.** or **P.M.**

We had lunch at 12:00 ____________.

9. Write 10 minutes after 10 o'clock on the clock face.

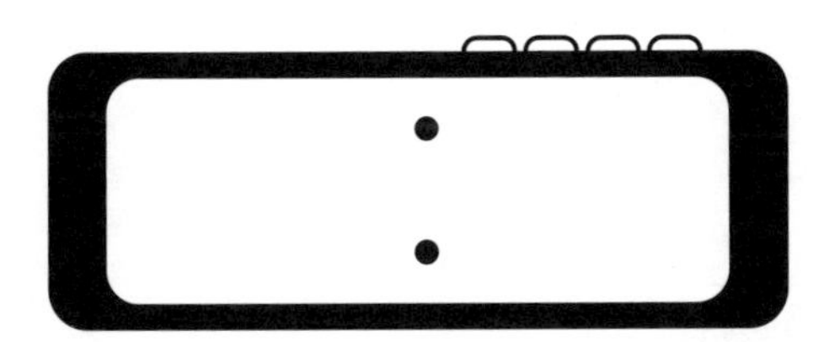

10. Draw the minute hand to show half past 2.

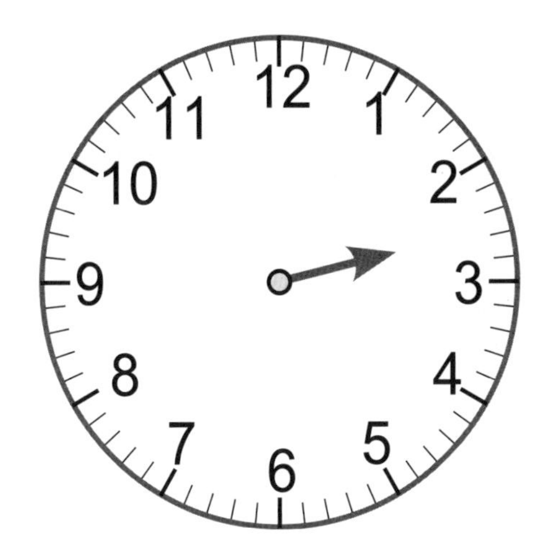

11. Fill in the blank.

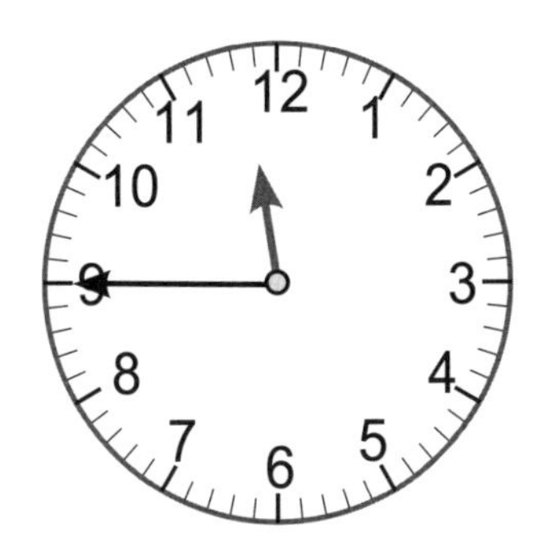

The time is ______________________ to noon.

12. Fill in the blank.

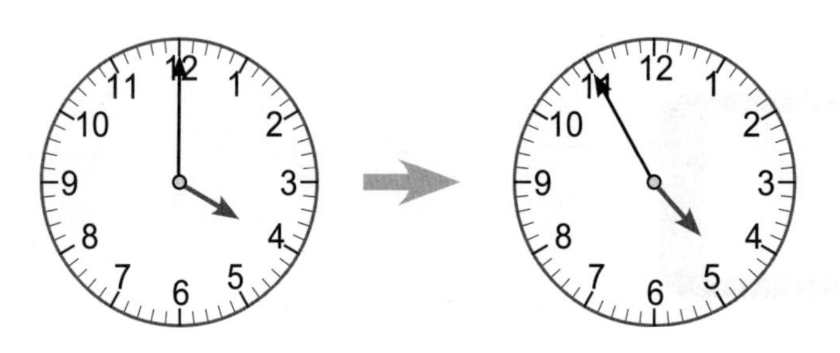

4:00 to ____________

13. Draw the missing hour hand on Clock B to show the time on Clock A.

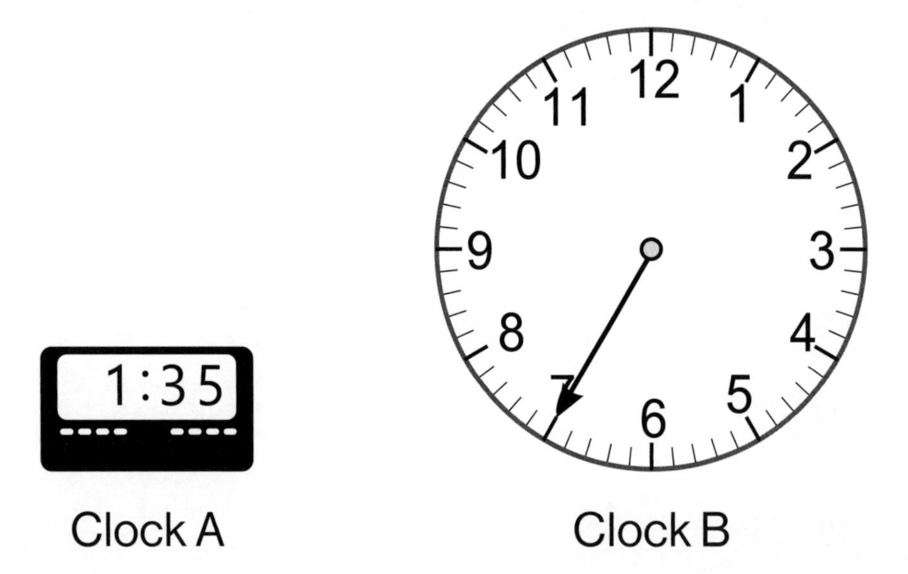

Clock A Clock B

14. What time is shown on the clock?

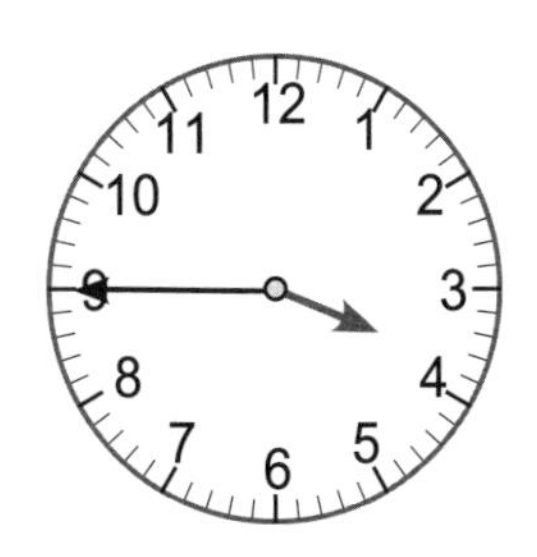

____________ minutes after ____________.

15. Lucy went to the library at 11:40 A.M.
Draw the hour and minute hands on the clock to show the time she went to the library.

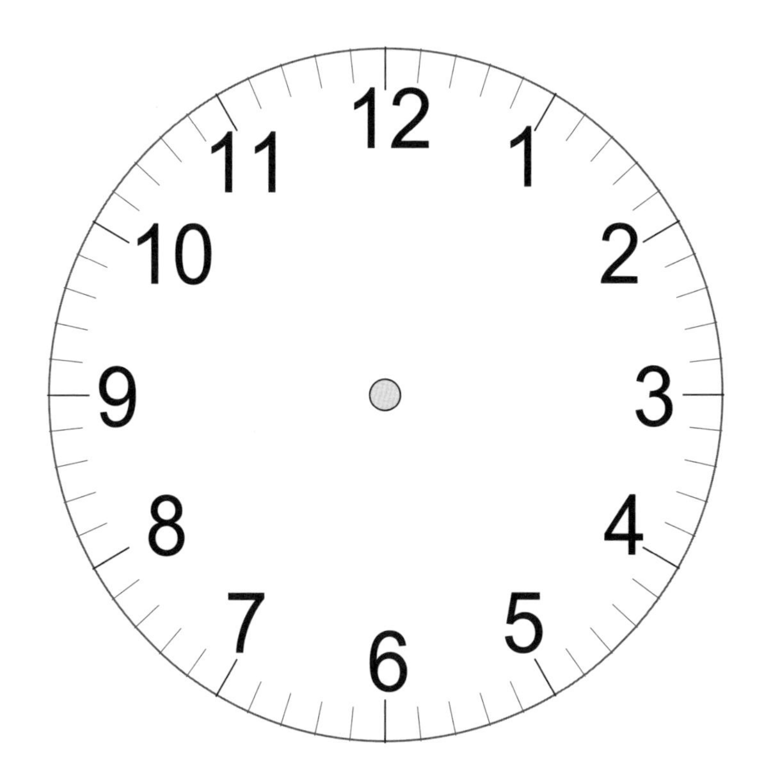

BLANK

Name: ______________________ Date: __________

Test A

30

Score

Unit 11 Graphs

Section A (2 points each)
Circle the correct option: **A**, **B**, **C**, or **D**.

The graph below shows the favorite flowers of a group of children.
Look at the graph and answer questions 1 to 5.

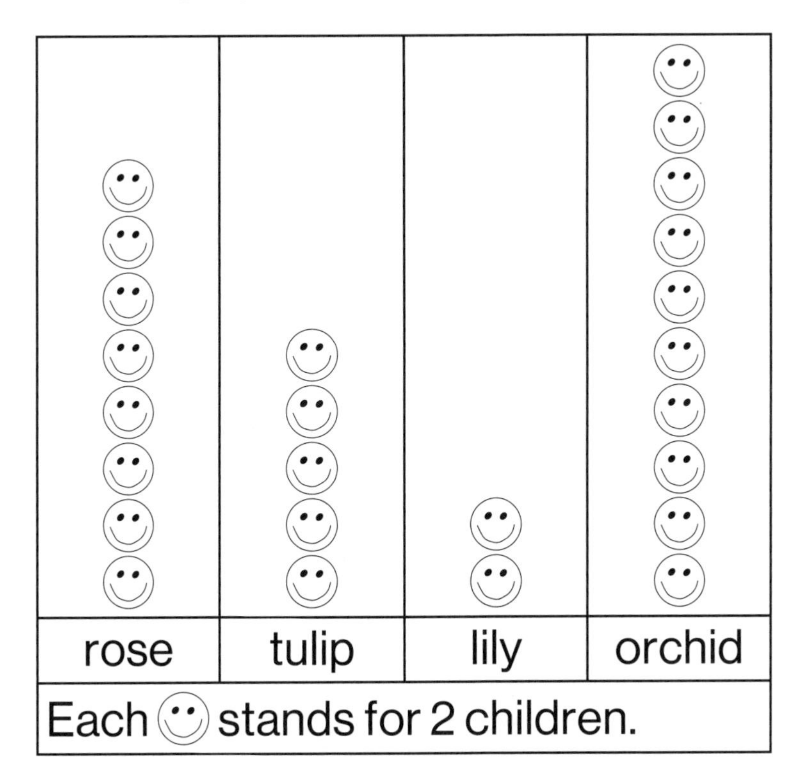

1. Which is the most popular flower?

A rose **B** tulip

C lily **D** orchid

2. How many children like roses?

A 14 **B** 16

C 18 **D** 8

3. How many children like tulips and lilies altogether?

A 7 **B** 12

C 14 **D** 16

4. How many more children like orchids than tulips?

A 5 **B** 6

C 10 **D** 12

5. How many children are there in the group altogether?

A 52 **B** 50

C 26 **D** 25

Section B (2 points each)

6.  stands for 24 ice-cream cones.

How many ice-cream cones does each  stand for?

Each  stands for ____________ ice-cream cones.

The graph below shows the number of different fish sold at a pet shop.
Look at the graph and answer question 7.

Number of fish sold at a pet shop

goldfish	 7 fish symbols
swordtail	 3 fish symbols
guppy	
clownfish	 5 fish symbols
 Each fish symbol stands for 3 fish.	

7. 3 more guppies than clownfish were sold at the shop.
Draw  in the graph to show the number of guppies sold.

The graph below shows the number of patients who visited the dentist in a week.
Answer questions 8 through 12.

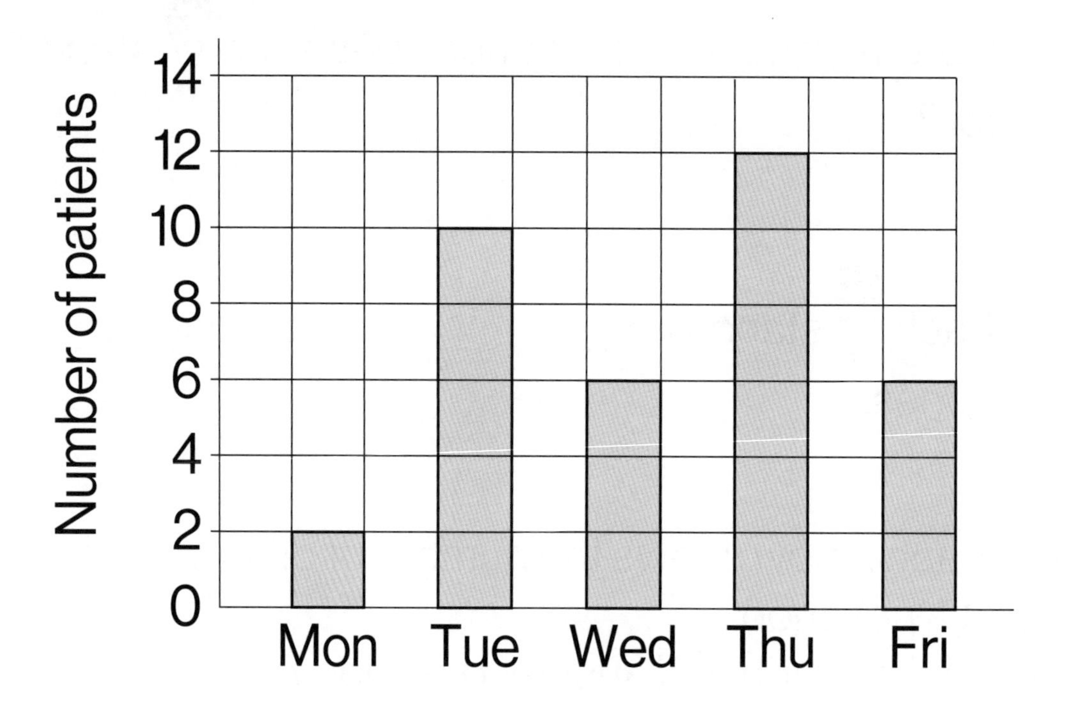

8. On which day was the dentist visited by the least number of patients?

9. On which two days did an equal number of patients visit the dentist?

10. How many more patients visited the dentist on Tuesday than on Wednesday?

11. Which day had twice as many patients as Friday?

12. How many patients visited the dentist on Thursday and Friday altogether?

Kenya plotted this graph to show the height of some trees in a park.

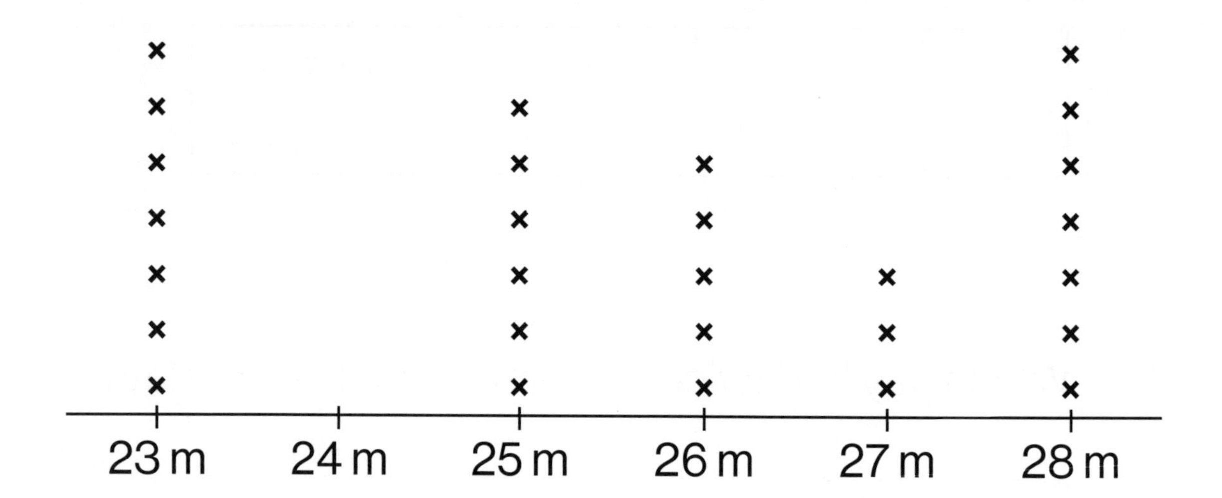

Look at the graph and answer questions 13 to 15.

13. How many trees are 25 m tall?

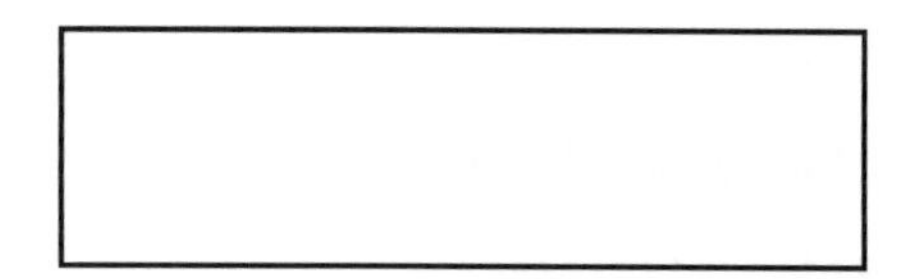

14. How many trees are less than 26 m tall?

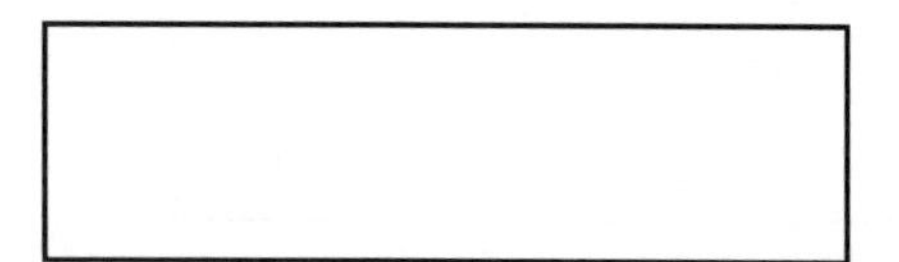

15. How many more trees are 28 m tall than 27 m tall?

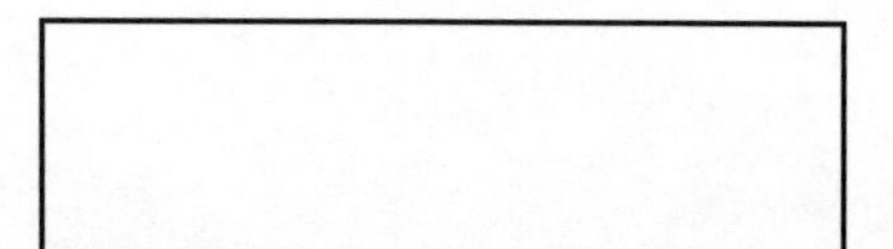

Name: ______________________ Date: __________

Test B

25 min 

Score
30

Unit 11 Graphs

Section A (2 points each)
Circle the correct option: **A**, **B**, **C**, or **D**.

The graph shows the favorite exercises of some students in a fitness club.
Look at the graph and answer questions 1 and 2.

jogging	cycling	swimming
Each stands for ? students.		

1. If 8 students like jogging, how many students does each stand for?

   A 16 **B** 2

   C 8 **D** 4

2. How many students are there in the club in all?

   A 9 **B** 13

   C 36 **D** 40

The graph below shows the number of marbles 4 students collected. Look at the graph and answer questions 3 to 5.

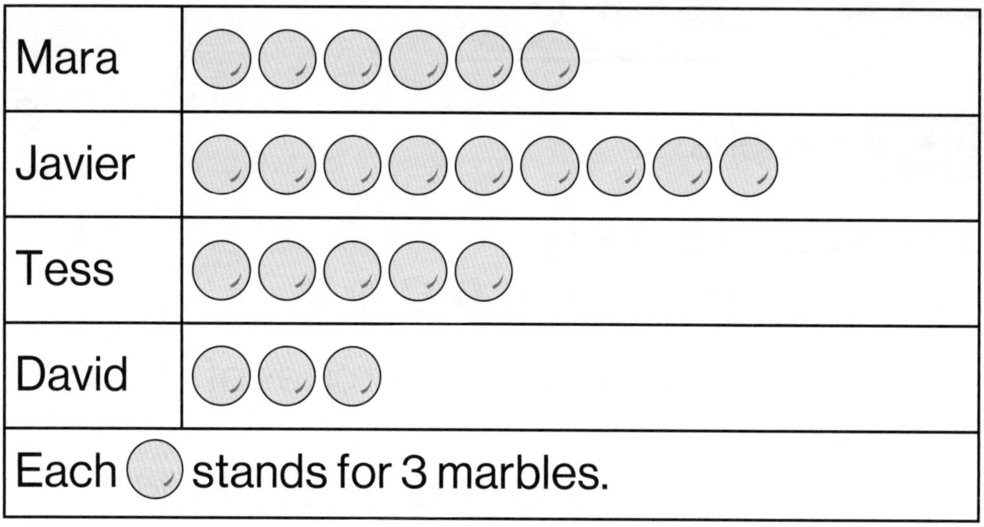

Mara	○○○○○○
Javier	○○○○○○○○○
Tess	○○○○○
David	○○○
Each ○ stands for 3 marbles.	

3. Who collected 18 marbles?

   A Mara **B** Javier

   C Tess **D** David

4. Which 2 students collected a total of 24 marbles?

   A Mara and Javier **B** Mara and Tess

   C Javier and Tess **D** Tess and David

5. How many marbles must Javier give to David so that both of them have the same number of marbles?

   A 9 **B** 6

   C 3 **D** 18

Section B (2 points each)

6. If stands for 10,

stands for ☐.

The graph below shows the number of stamps collected by 5 students.
Look at the graph and answer questions 7 to 9.

Number of stamps collected by 5 students

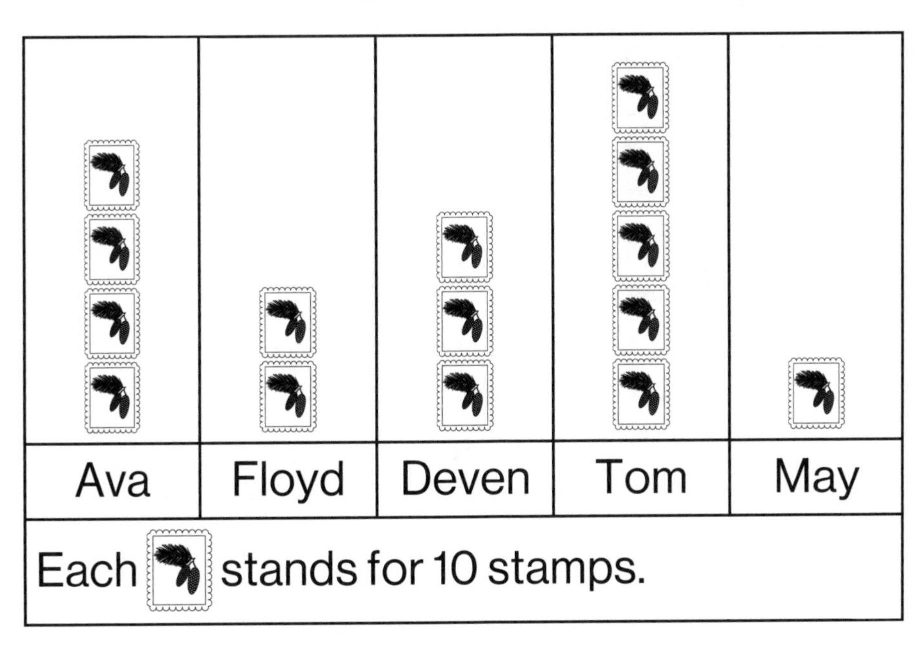

7. Who collected fewer stamps than Floyd?

☐

8. How many children collected more than 20 stamps?

9. How many more stamps did Tom collect than May?

Henry recorded the number of vehicles in a parking lot in the picture graph as shown below.
Use the information to answer questions 10 to 12.

bus	⊛ ⊛ ⊛ ⊛
motorbike	⊛ ⊛ ⊛ ⊛ ⊛ ⊛
car	⊛ ⊛ ⊛ ⊛ ⊛ ⊛ ⊛ ⊛
van	⊛ ⊛ ⊛ ⊛ ⊛ ⊛ ⊛ ⊛ ⊛ ⊛
truck	⊛ ⊛
Each ⊛ stands for 3 vehicles.	

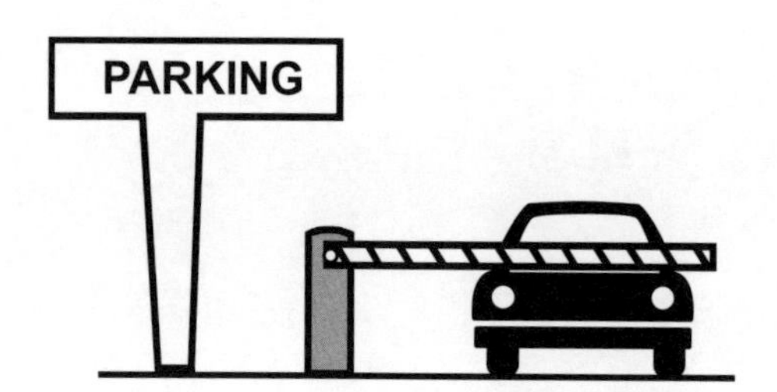

10. Shade in the bar graph below to show the correct number of cars.

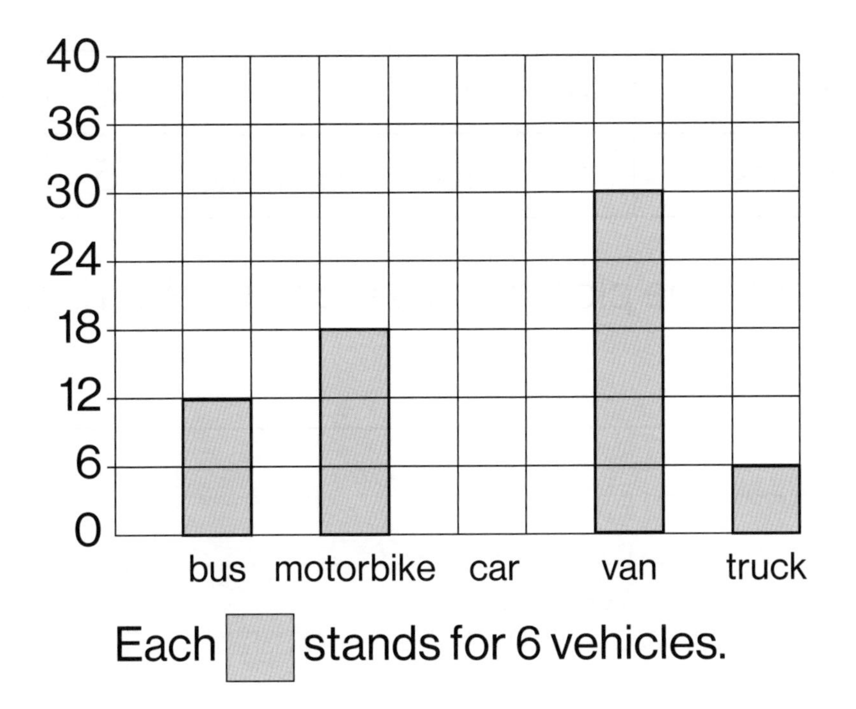

Each [ ] stands for 6 vehicles.

11. How many vehicles are there in the parking lot altogether?

[ ]

12. If the parking fee for each vehicle is $5, how much money was collected from vans and trucks?

$ [ ]

This table shows the height of some tulip plants in a garden.

Height in inches	4	5	6	7	8	9
Number of plants	5	10	15	12	5	8

13. Complete the line plot to show the data given in the table.

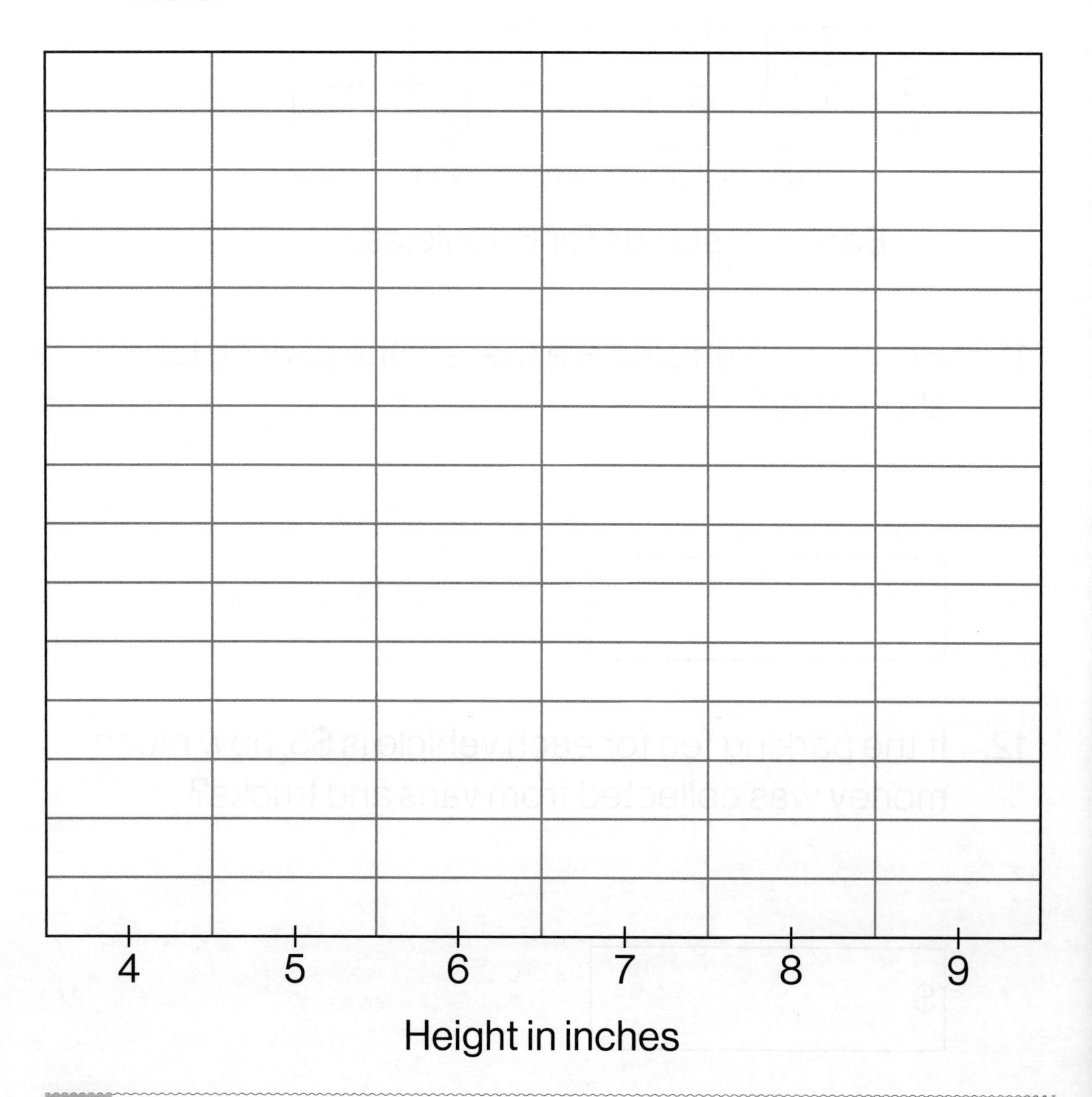

14. What is the most common height of the tulip plants?

15. How many plants are less than 6 inches tall?

BLANK

Name: ______________________ Date: __________

Test A

30
Score

Unit 12 Geometry

Section A (2 points each)
Circle the correct option: **A**, **B**, **C**, or **D**.

1. Which solid has only 1 flat surface and 1 curved surface?

A 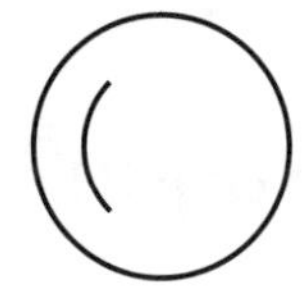

B 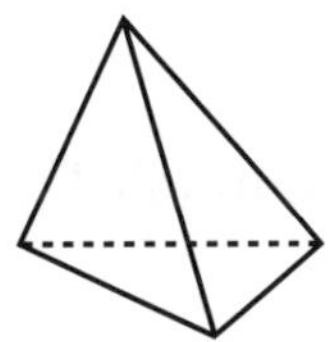

C 

D 

2. Which figure is formed by a half circle and a rectangle?

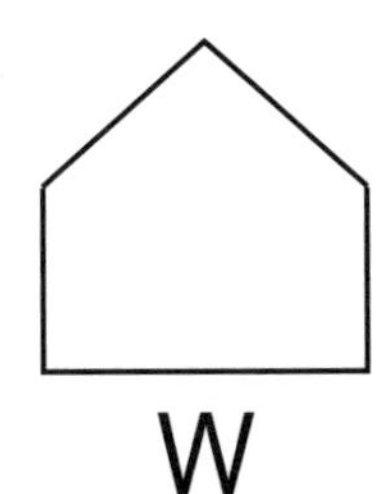
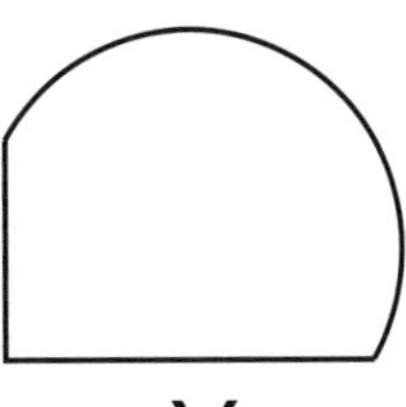
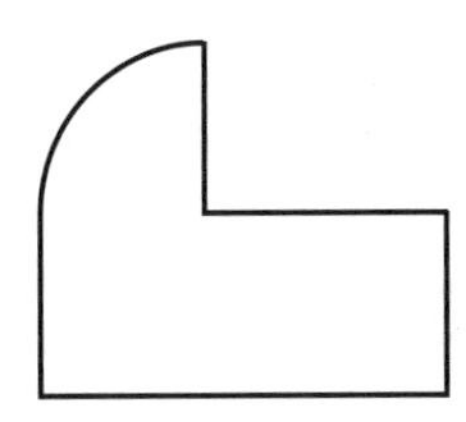

W X Y Z

A W

B X

C Y

D Z

3. What is the missing shape in the pattern below?

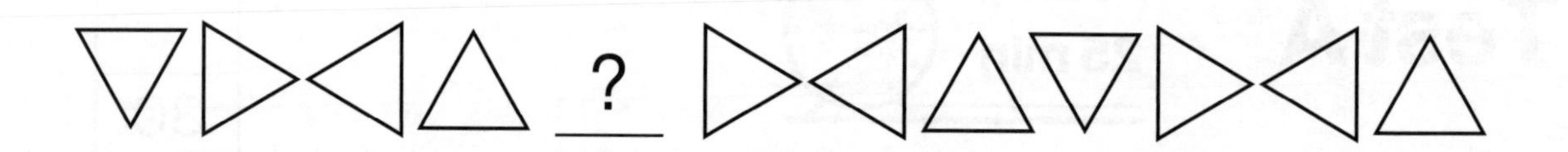

A 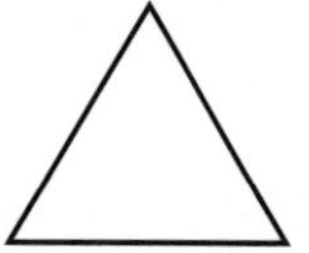

B 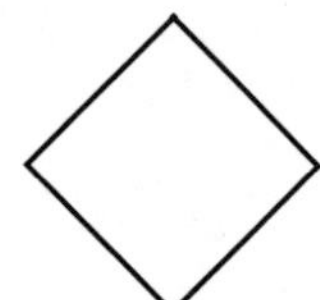

C 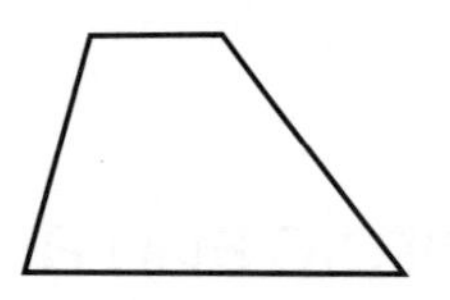

D 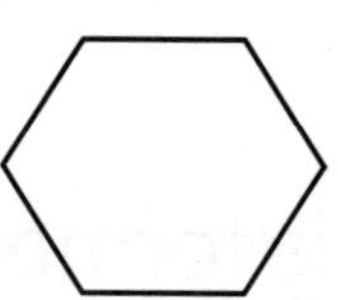

4. Which of the following figures is a hexagon?

A 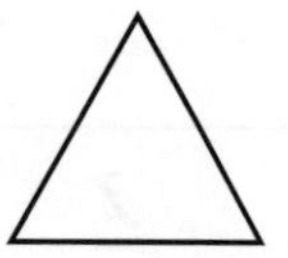

B 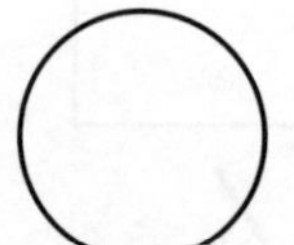

C 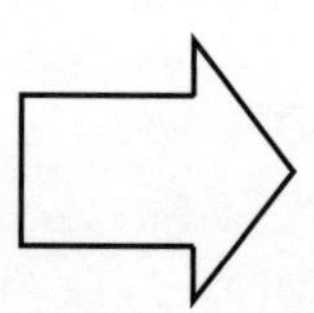

D 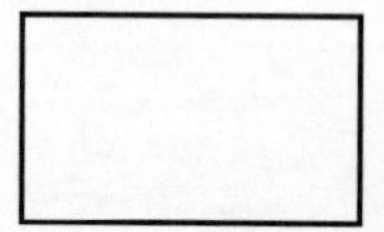

5. Which figure has no angle in it?

A

B

C

D

Section B (2 points each)

6. Color the solid that has flat surfaces only.

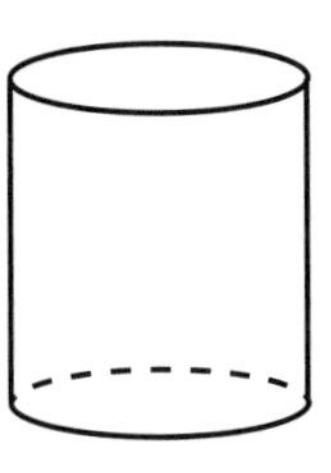

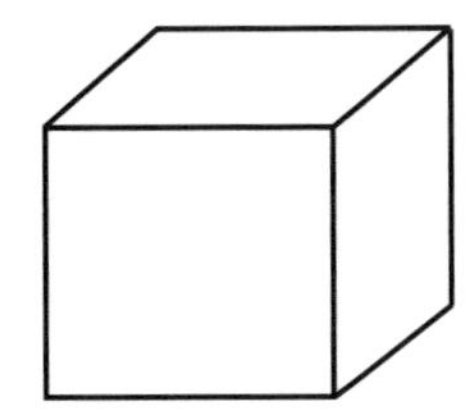

7. Color the shape that has both straight lines and curves.

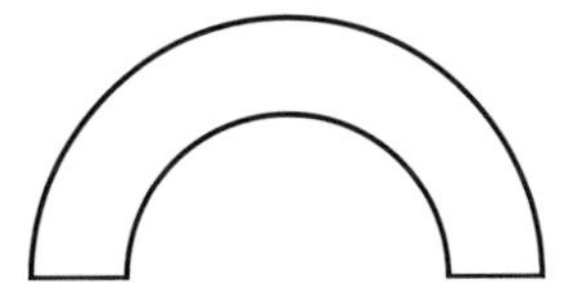

8. Name the 3 shapes that make up the figure below.

9. How many angles are in the figure above?

Answer: ______________________

10. There are ____________ quarter circles in the figure below.

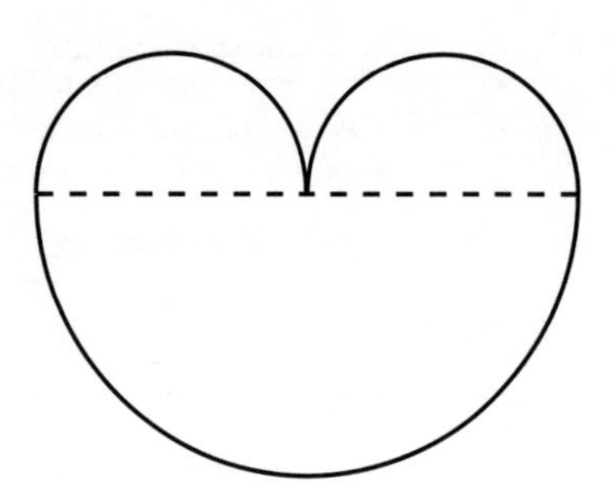

11. The figure below is formed by ____________

straight lines and ____________ curves.

12. Color the solid formed by straight lines only.

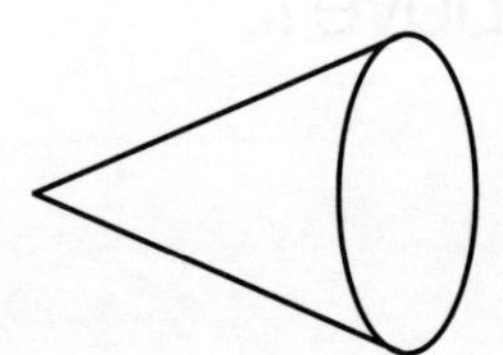

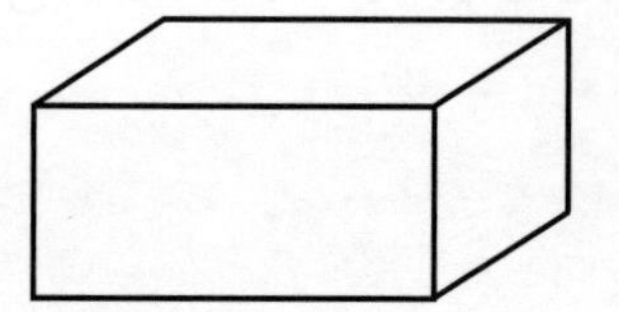

13. Look at the pattern.
Circle the solid that is missing.

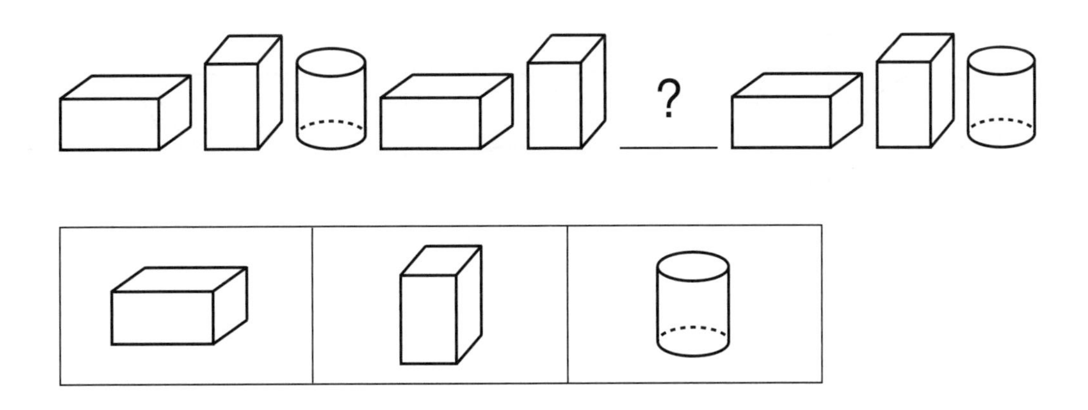

14. Copy the arrow using the grid on the right.

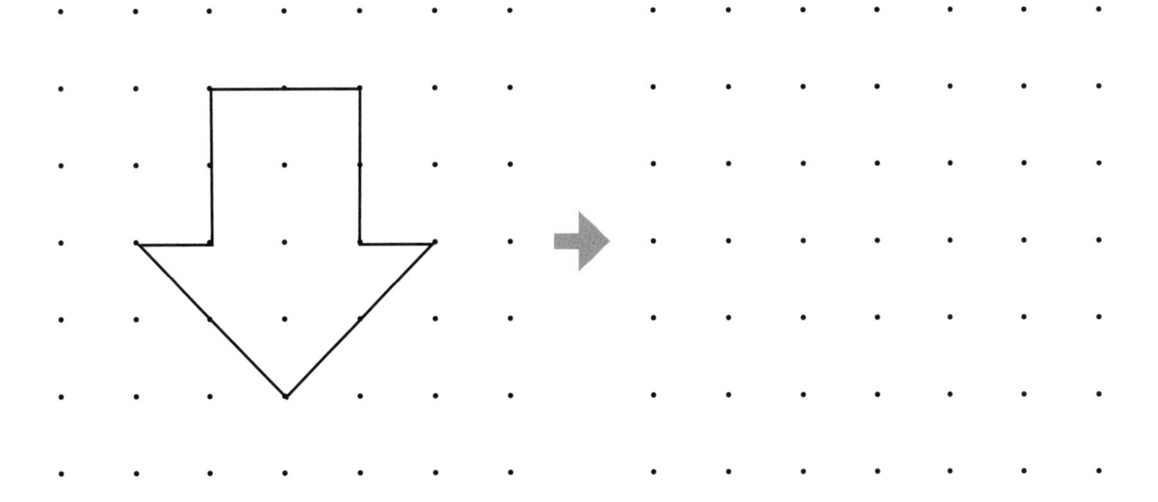

15. Look at the pattern.
Draw the shape that comes next.

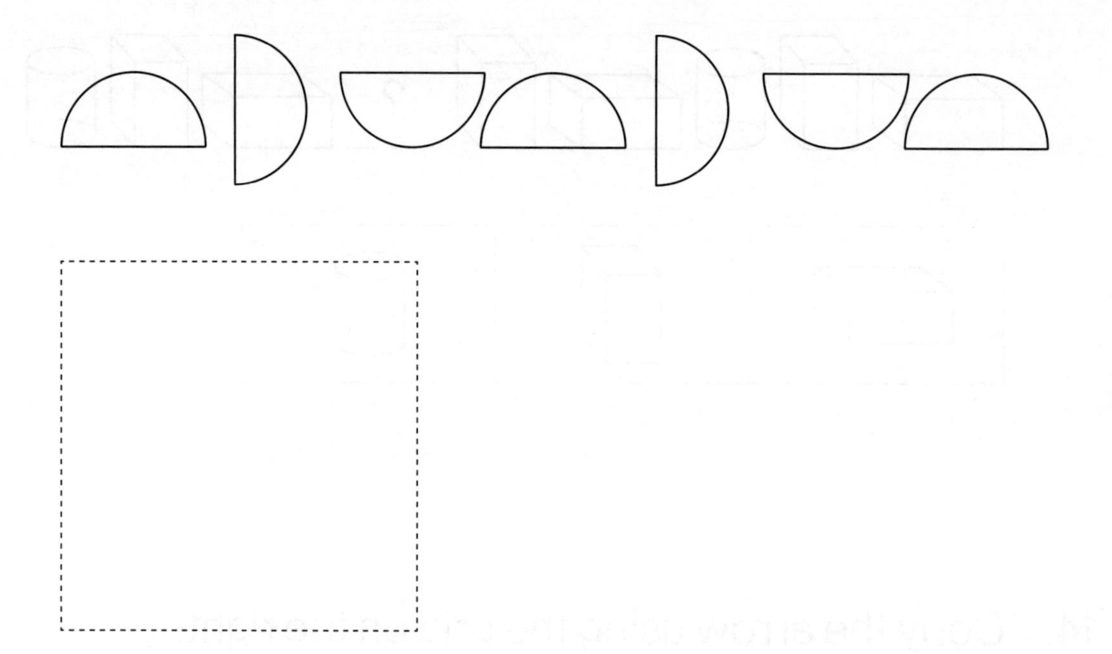

Name: ______________________________ Date: __________

Test B

30
Score

Unit 12 Geometry

Section A (2 points each)
Circle the correct option: **A**, **B**, **C**, or **D**.

1. Which solid below has 5 flat surfaces?

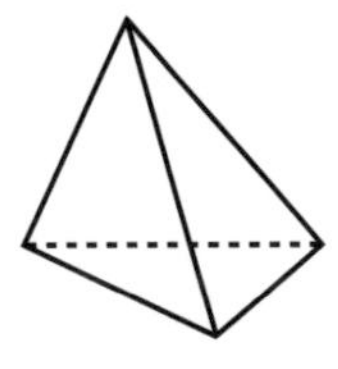
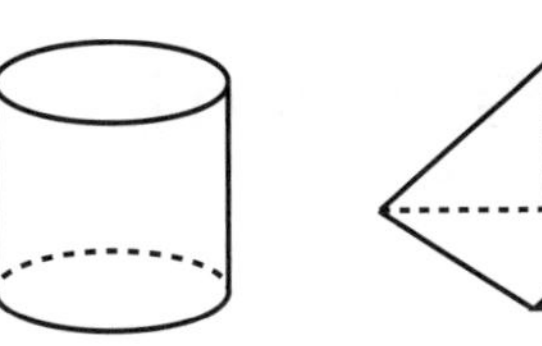
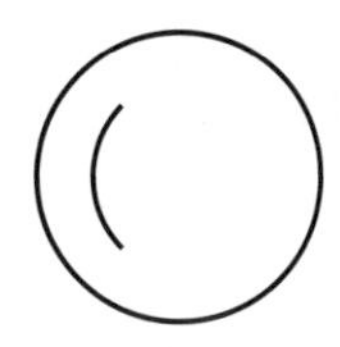

A B C D

A Solid A **B** Solid B

C Solid C **D** Solid D

2. Look at the pattern.
   What shape comes next?

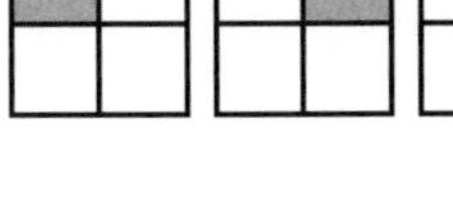
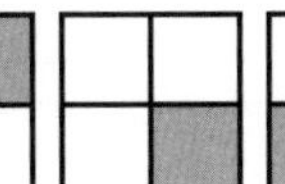
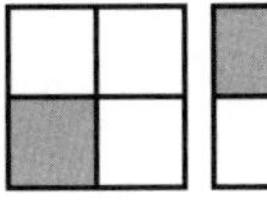

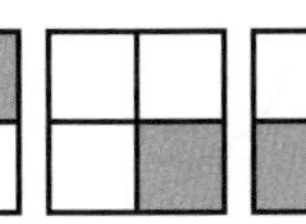
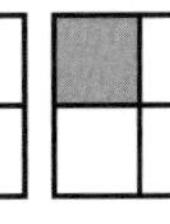

? ______

A 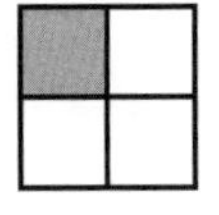 **B** 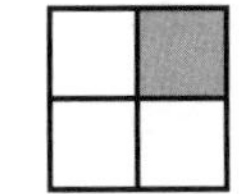

C 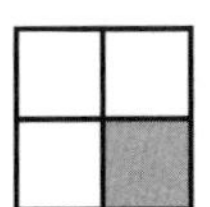 **D** 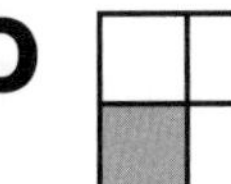

Look at the figures below. Answer questions 3 and 4.

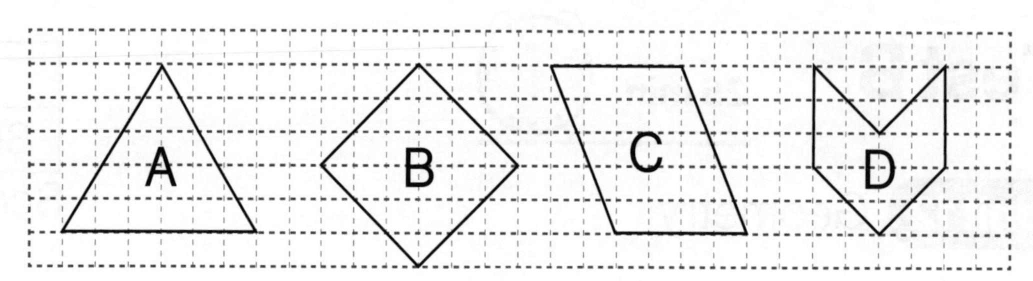

3. Which figures are quadrilaterals?

A A and B **B** B and C

C C and D **D** A and D

4. What is figure D called?

A Quadrilateral **B** Pentagon

C Hexagon **D** Octagon

5. How many angles are there in the figure?

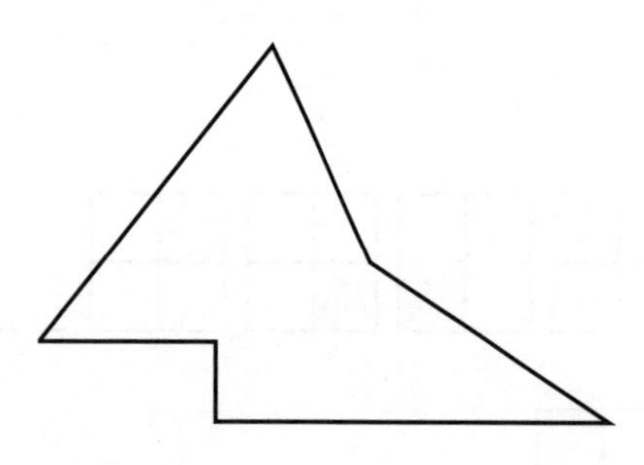

A 5 **B** 6

C 3 **D** 4

Section B (2 points each)

6. How many flat surface(s) is/are there in the solid below?

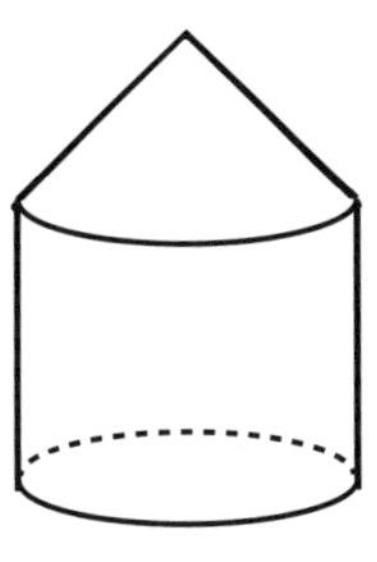

Answer: ______________________________

7. How many quarter circles can you find in the figure below?

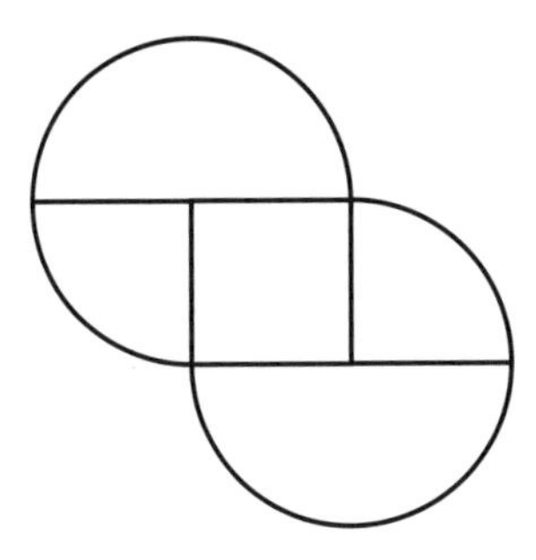

Answer: ______________________________

8. The figure below is formed by 4 shapes.
Draw 3 straight lines to show how it is formed by the 4 shapes.

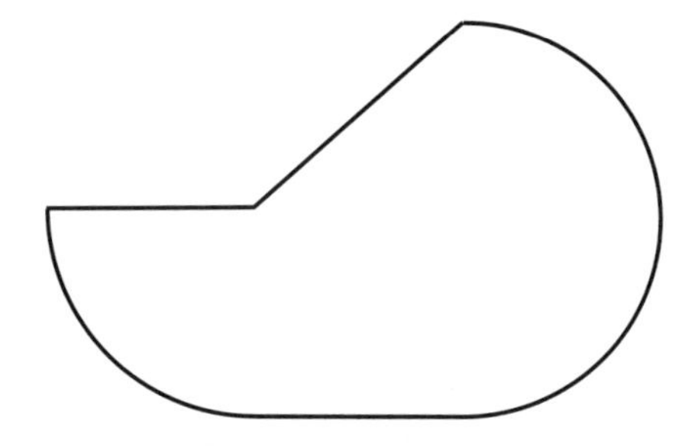

9. Copy the figure.

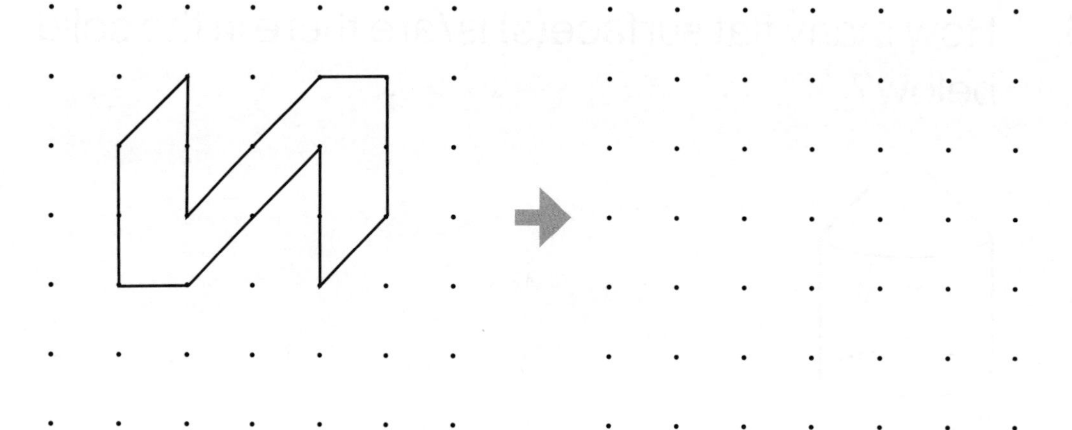

10. How many angles does the figure above have?

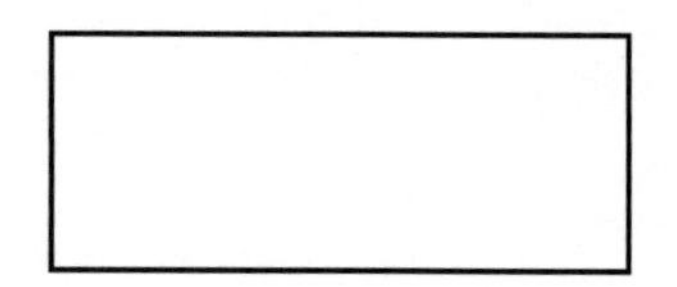

11. What kind of polygon is the figure?

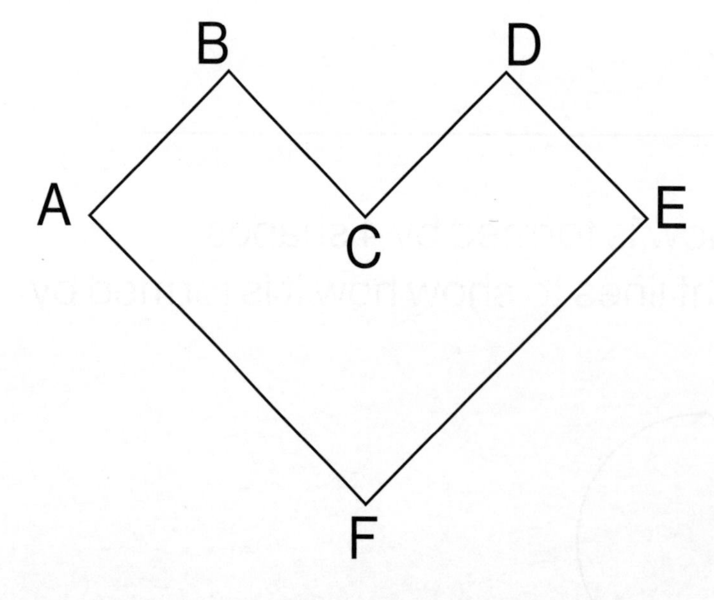

Answer: ______________________

12. Draw 3 straight lines in the figure to get 4 equal triangles.

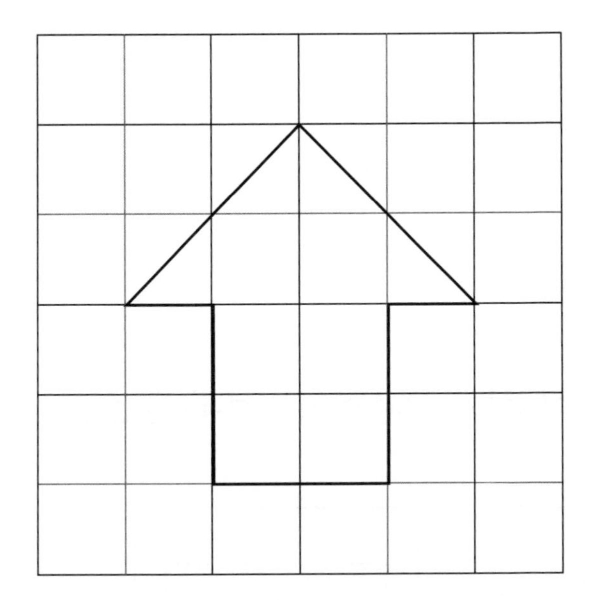

13. Circle the shape that comes next.

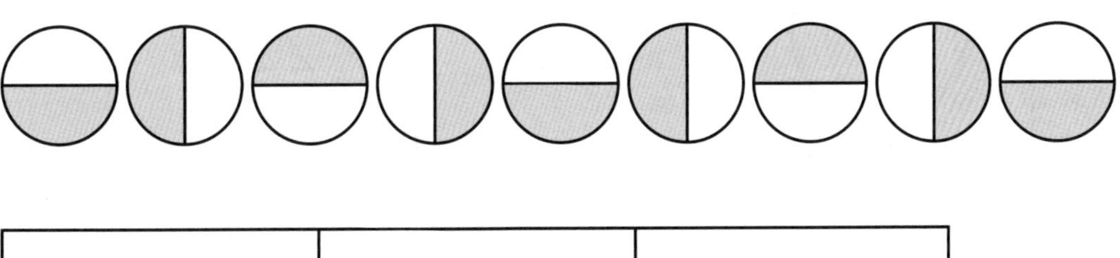

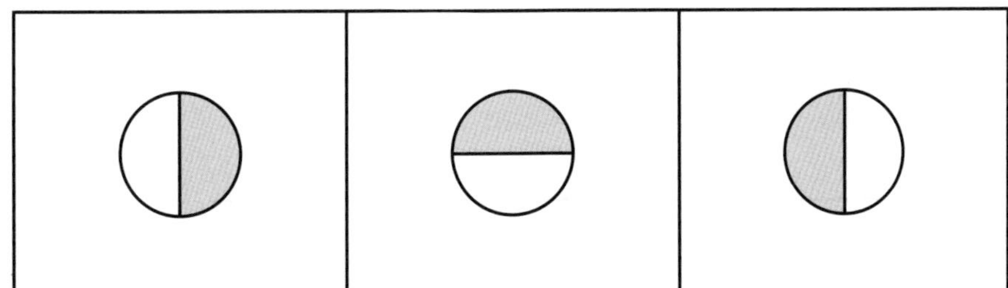

14. Draw the shape that comes next.

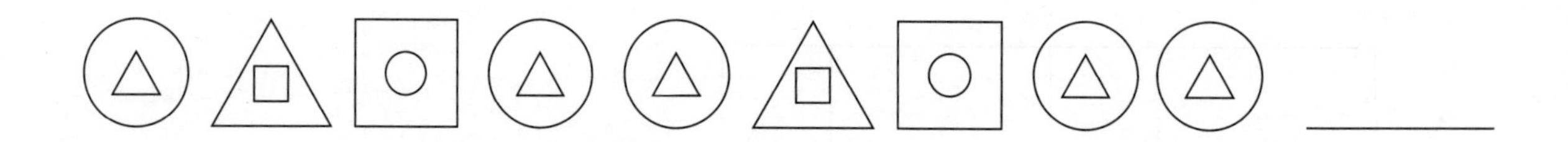

15. Look at the pattern.
Draw the missing symbol in the dotted-line box.

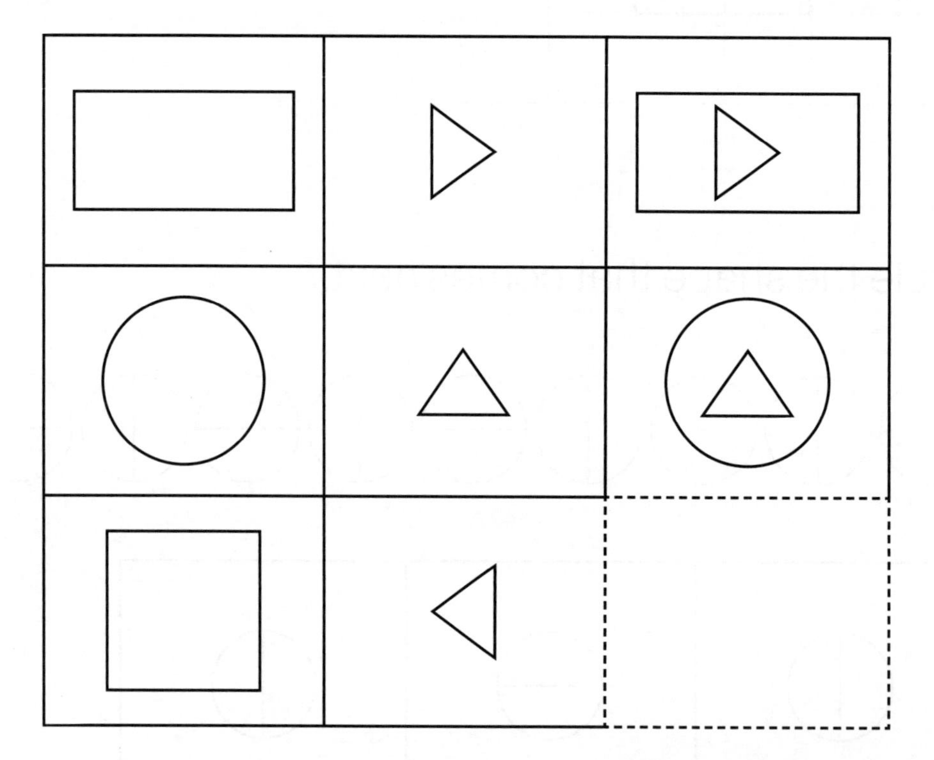

Name: ______________________ Date: __________

Test A

80

Score

Year-End Assessment

Section A (2 points each)
Circle the correct option: **A**, **B**, **C**, or **D**.

1. 5 hundreds and 9 ones is the same as __?__.

   A 950 **B** 905

   C 590 **D** 509

2. The missing number in the pattern is __?__.

   A 704 **B** 734

   C 744 **D** 754

3. 239 is __?__ tens more than 209.

   A 30 **B** 2

   C 3 **D** 20

4. Subtract ___?___ from 166.
The answer is 100.

A 66 **B** 156

C 161 **D** 266

5. Each  stands for 10 stickers.
stands for ___?___ stickers.

A 105 **B** 60

C 50 **D** 15

6. 6 dollars and 9 cents is the same as ___?___.

A $69 **B** $6.09

C $6.90 **D** $0.69

7. What is the missing sign?
$21 \div 3 = 3 \boxed{?} 4$

A + **B** –

C × **D** ÷

8. What is the time shown?

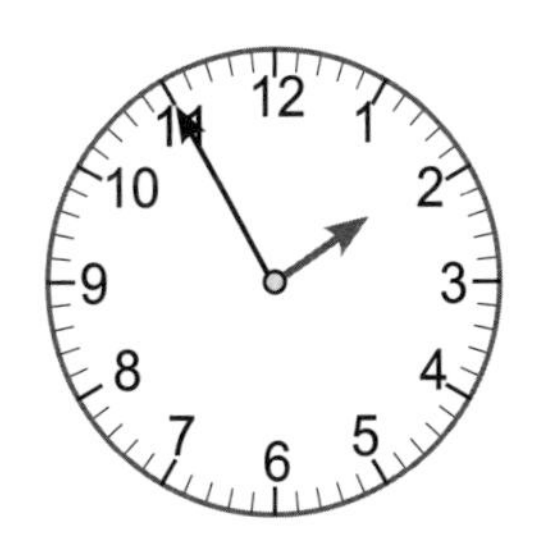

A 11:10 **B** 2:11

C 1:55 **D** 2:55

9. What fraction of the figure is shaded?

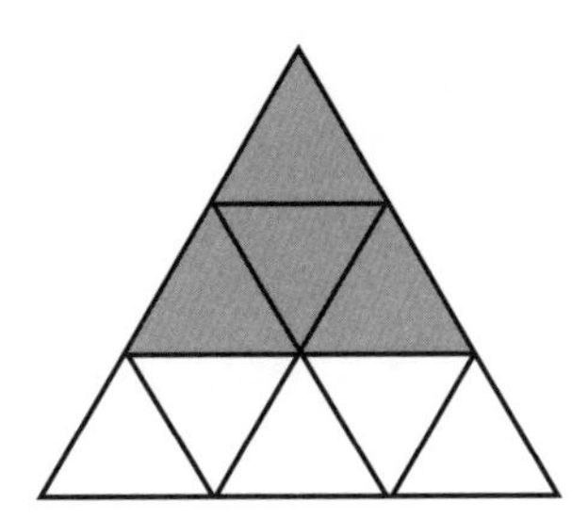

A $\frac{4}{9}$ **B** $\frac{4}{5}$ **C** $\frac{5}{9}$ **D** $\frac{1}{3}$

10. $\frac{3}{10}$ and ___?___ make 1 whole.

A $\frac{2}{10}$ **B** $\frac{5}{10}$ **C** $\frac{6}{10}$ **D** $\frac{7}{10}$

11. Which letter is made up of both straight line(s) and curve(s)?

S M A R T

A S **B** M

C A **D** R

12. How many △ are needed to form the shape below?

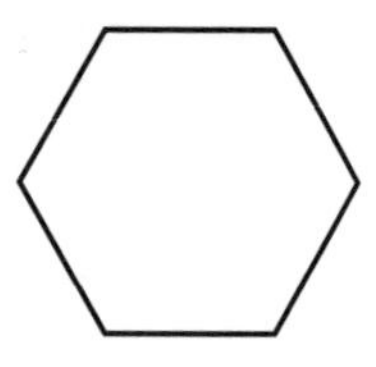

A 8 **B** 6

C 5 **D** 4

13. This figure is a __?__.

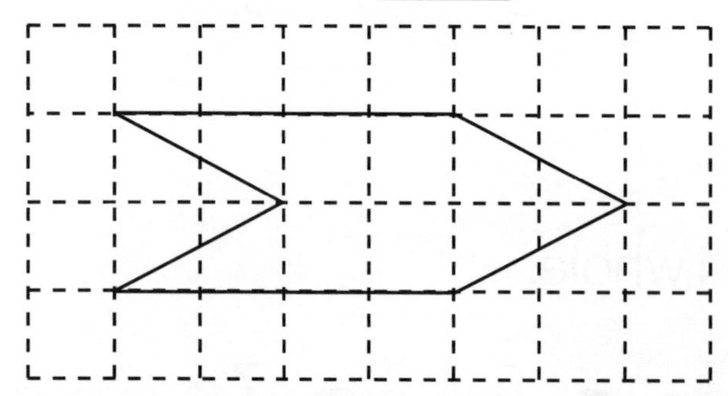

A Quadrilateral **B** Pentagon

C Hexagon **D** Octagon

The picture graph shows the number of students from 3 classes in the library during recess.

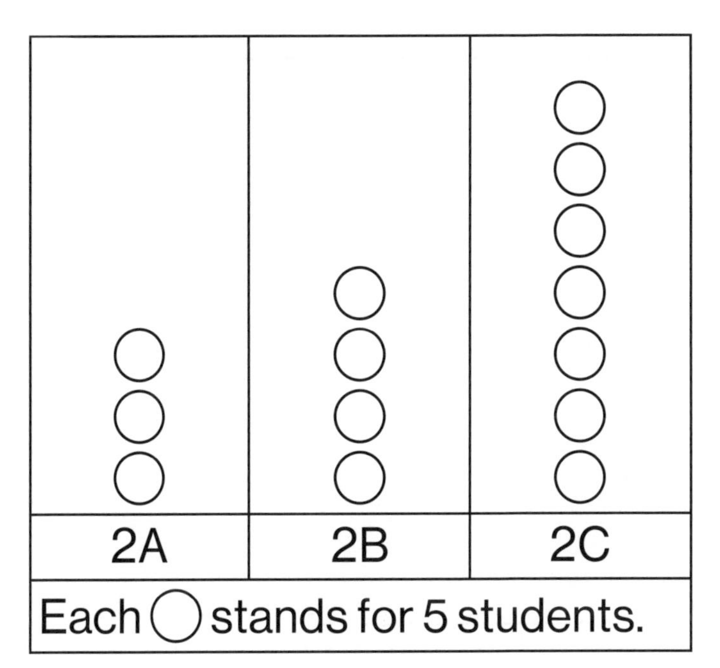

14. There are __?__ more students from 2C than from 2B in the library during recess.

A 15 **B** 10

C 3 **D** 8

15. A total of 50 students in the library are from __?__.

A 2A and 2B **B** 2B and 2C

C 2A and 2C **D** 2A, 2B and 2C

Section B (2 points each)

16. Write in words.

590 [ ]

17. The time shown is ____________.

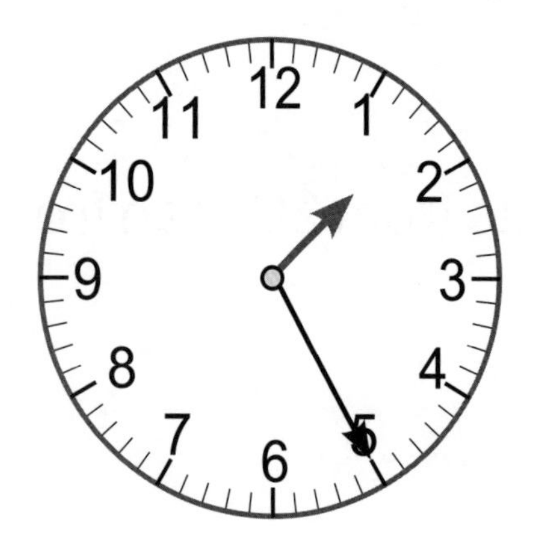

18. 3 tens and ____________ ones make 100.

19.

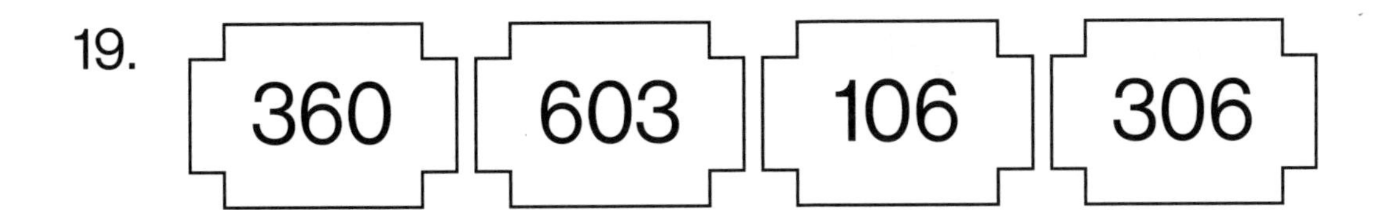

Add the greatest number and the smallest number from above.

The answer is ☐.

20. 6 books are put in each drawer.

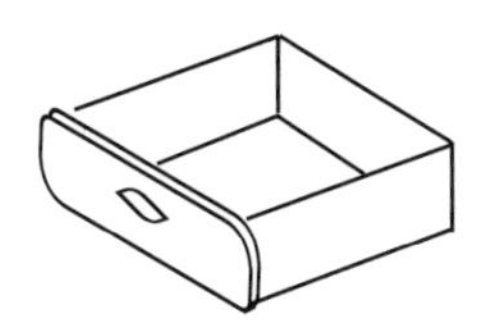
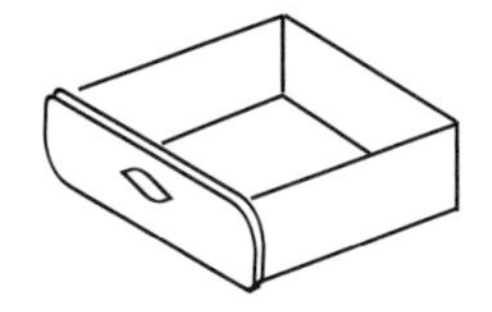
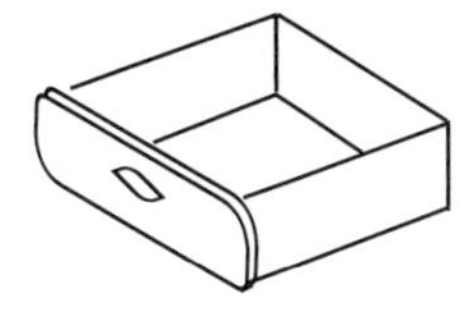
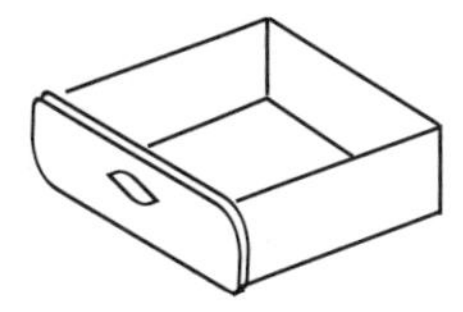

There are ______________ books in 4 drawers.

21. Write the missing digit.

$$\begin{array}{r} 2\ 2\ 6 \\ +\ 3\ 3\ \square \\ \hline 5\ 6\ 3 \\ \hline \end{array}$$

22. Write the missing digit.

$$\begin{array}{r} 5\ 0\ 8 \\ -\ 1\ \square\ 3 \\ \hline 3\ 7\ 5 \\ \hline \end{array}$$

23. If ♡ stands for 6, ♡ ♡ ♡ ♡ ♡ stands for

______________.

24. Arrange the fractions in order.
Begin with the smallest.

$$\frac{1}{2}, \quad \frac{1}{4}, \quad \frac{1}{11}$$

25. Color to show $\frac{1}{4}$.

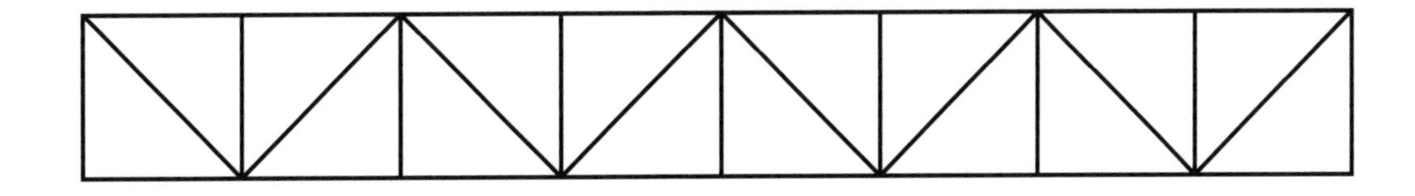

26. Look at the solid object.

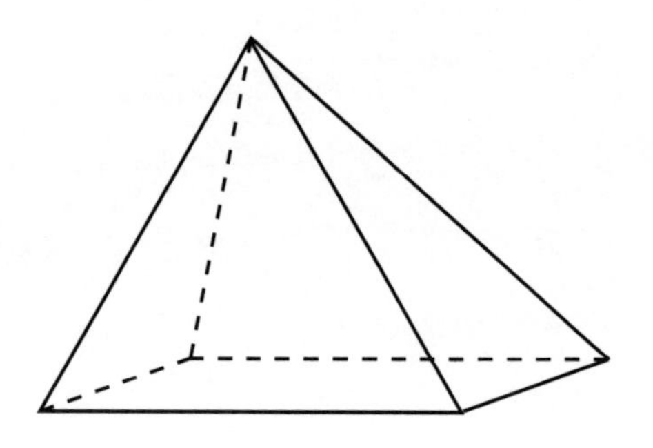

There are ____________ flat surfaces.

27. How many angles are there inside the figure?

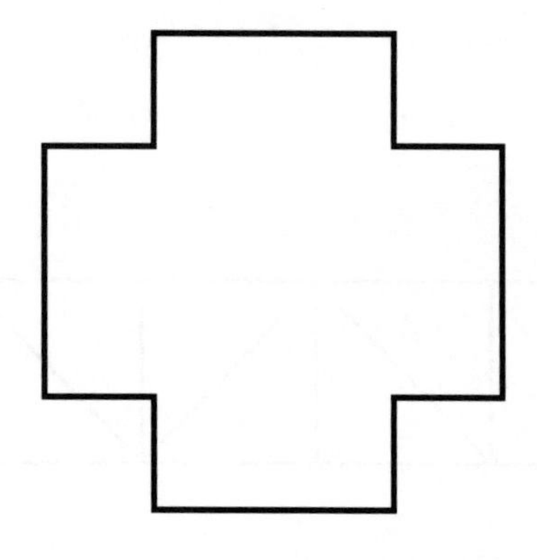

Answer: ____________________________

The graph shows the length of 4 poles.

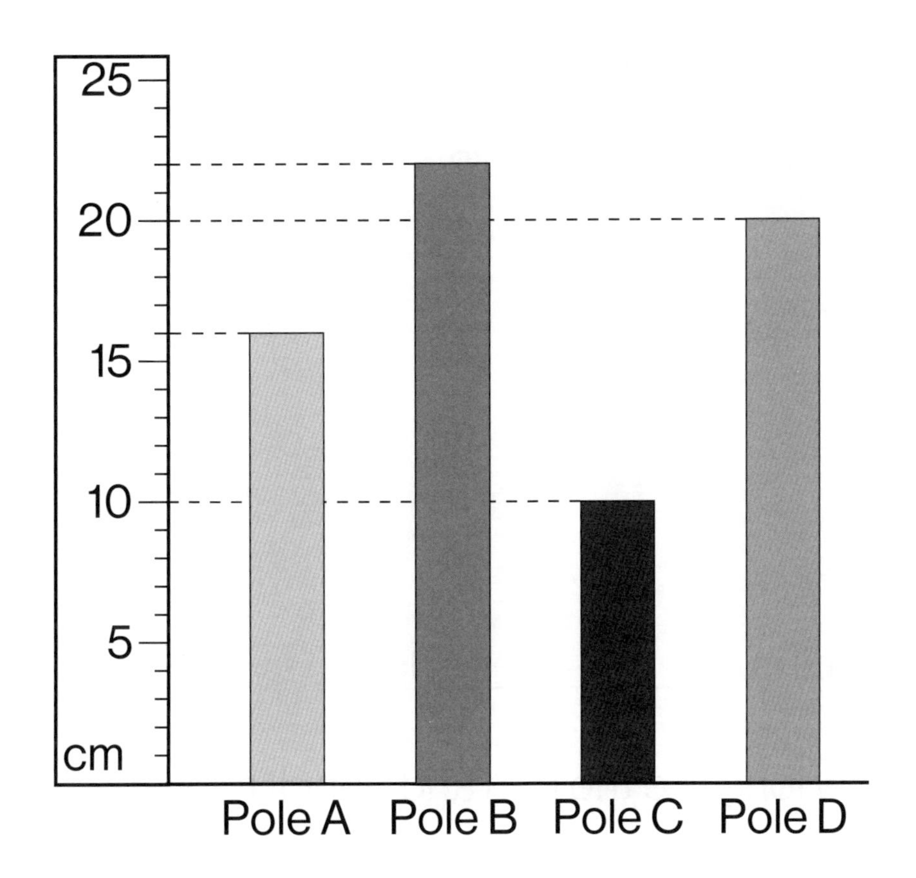

28. Pole ___________ is the longest.

29. Pole ___________ is 4 cm shorter than Pole D.

30. The total length of two poles is 26 cm.

    The two poles are ___________ and ___________.

Section C (4 points each)

31. A pen is 22 cm long.
It is 5 cm longer than a pencil.

a) How long is the pencil?

The pencil is ______________ long.

b) How long are the pen and pencil together?

The pen and pencil are ______________ together.

32. A toy car costs $2.60.
A toy boat costs $3.80 more than the toy car.

a) What is the cost of the toy boat?

The cost of the toy boat is ____________.

b) What is the total cost of the two toys?

The total cost of the two toys is ____________.

33. There were 760 apples in a crate.
Josh sold 35 apples on Saturday and 462 apples on Sunday.

a) How many apples did he sell both days?

He sold ______________.

b) How many apples were left in the crate?

______________ were left in the crate.

34. Mary has 6 quarters, 7 dimes, 4 nickels, and 8 pennies.
How much money does she have?

35. Darren cut a cake into 8 equal slices.
He ate 3 slices.

a) What fraction of the cake did he eat?

He ate ____________ of the cake.

b) What fraction of the cake was left?

____________ of the cake was left.

Extra Credit

1. Brandon has 4 ropes A, B, C, and D.
A is shorter than B.
C is shorter than A but longer than D.

The longest rope is ______________.

2. A pizza was cut into quarters as shown below.
Jolene took one piece and ate $\frac{1}{3}$ of it.

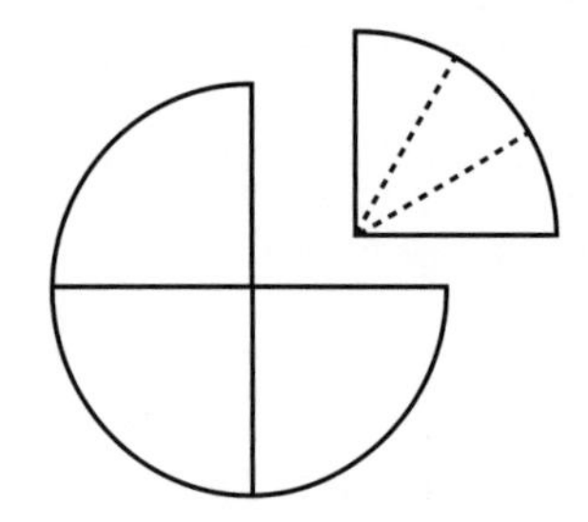

a) What fraction of the pizza did she eat?

She ate ______________ of the pizza.

b) What fraction of the pizza was left?

______________ of the pizza was left.

Name: ______________________________ Date: __________

Test B

80

Score

Year-End Assessment

Section A (2 points each)
Circle the correct option: **A**, **B**, **C**, or **D**.

1. Which letter is made up of both straight line(s) and curve(s)?

A N G L E S

A A **B** N

C G **D** S

2. A teacher's table is about __?__ high.

A 8 cm **B** 80 cm

C 8 m **D** 80 m

3. A cake was cut into 8 equal pieces.
Autumn ate 3 pieces.
What fraction of the cake was left?

A $\frac{2}{8}$ **B** $\frac{3}{8}$ **C** $\frac{5}{8}$ **D** $\frac{7}{8}$

4. Niko put 5 apples in a bag.
How many apples are there in 9 bags?

A 14 **B** 40

C 45 **D** 59

5. What is the missing number?

$35 \div 5 = 21 \div \boxed{?}$

A 1 **B** 5

C 3 **D** 7

6. Which of the following is correct?

A $\frac{1}{6} > \frac{1}{7}$ **B** $\frac{1}{3} = \frac{2}{3}$

C $\frac{1}{4} < \frac{1}{5}$ **D** $\frac{1}{11} > \frac{1}{2}$

7. Look at the given figure.

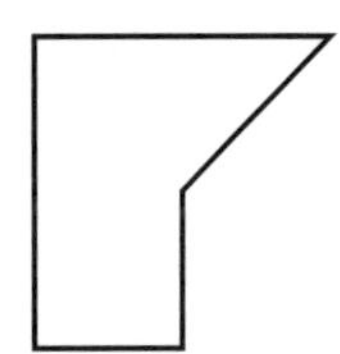

Which is the missing part needed to make it a square?

A  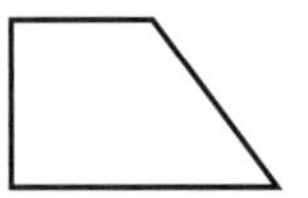

B 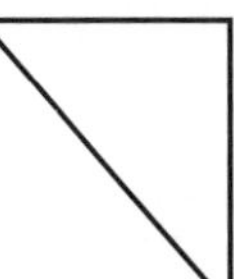

C 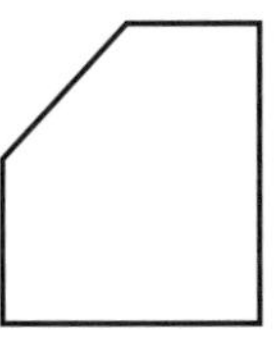

D 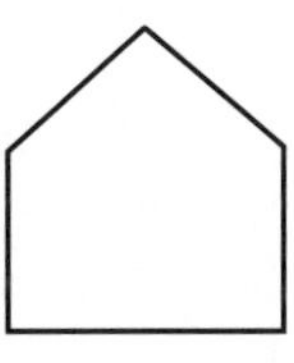

8. The figure has __?__ flat surfaces.

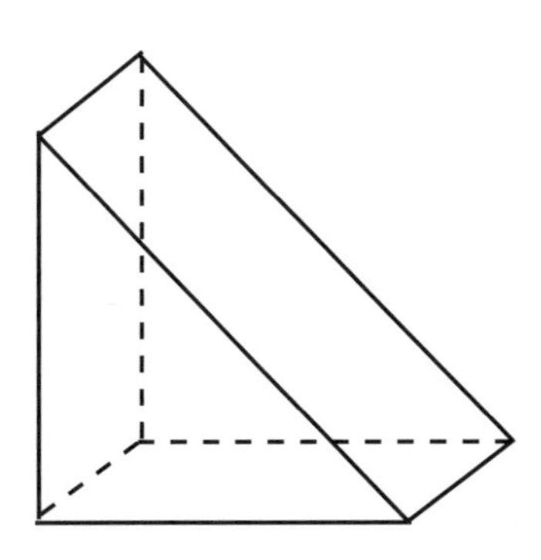

A 8

B 7

C 6

D 5

9. 40 is equal to 8 groups of ___?___.

A 8 **B** 6

C 5 **D** 4

10. What fraction of the figure is shaded?

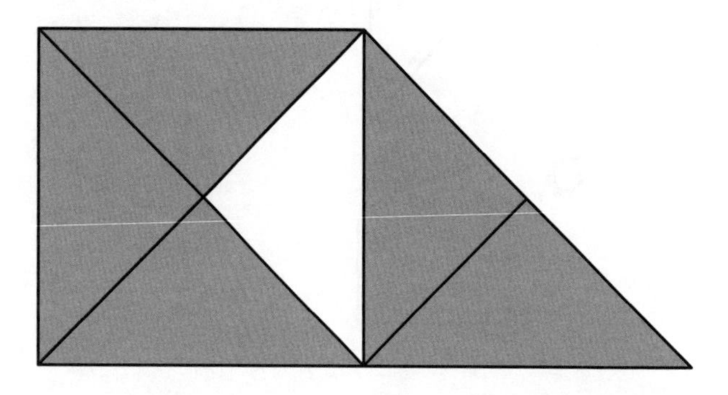

A $\frac{4}{6}$ **B** $\frac{2}{6}$ **C** $\frac{2}{4}$ **D** $\frac{5}{6}$

11. Which angle is the smallest?

A

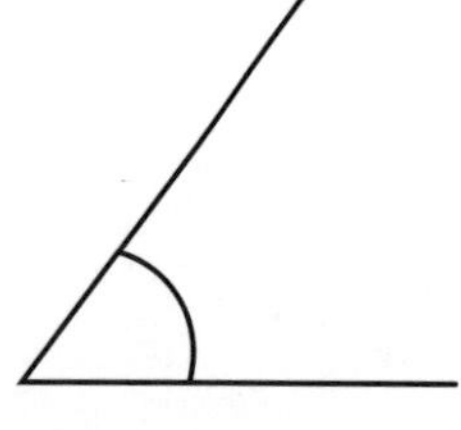

B

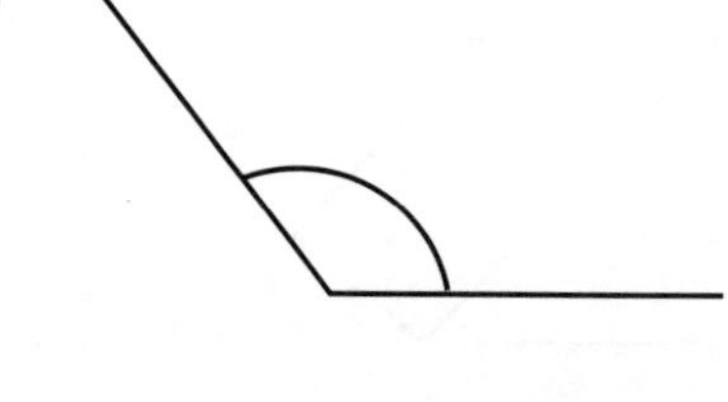

C

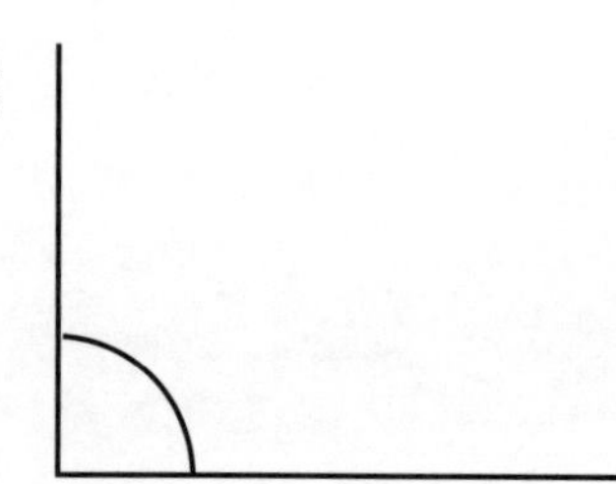

D

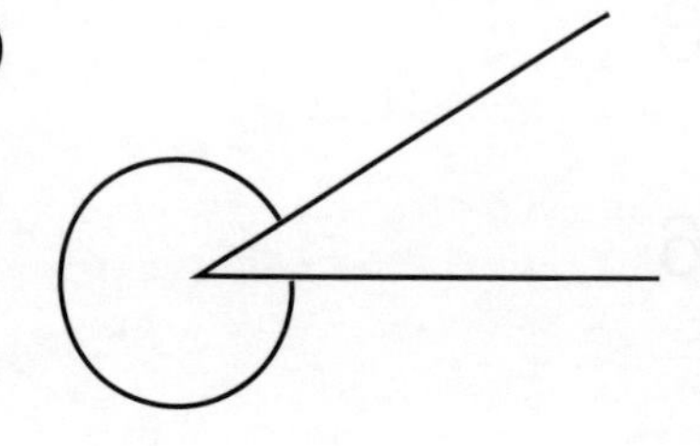

12. Look at the figure below.
How many angles are there inside the figure?

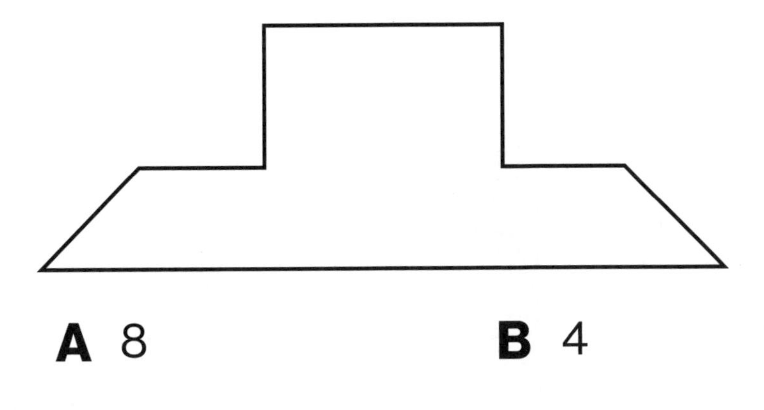

A 8 **B** 4

C 2 **D** 12

13. This figure is a ___?___.

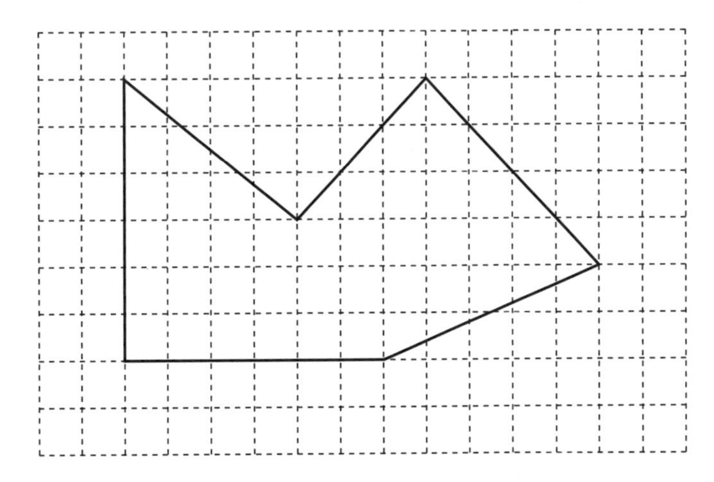

A Quadrilateral **B** Pentagon

C Hexagon **D** Octagon

The graph shows the favorite exercises of the students in Grade 2.

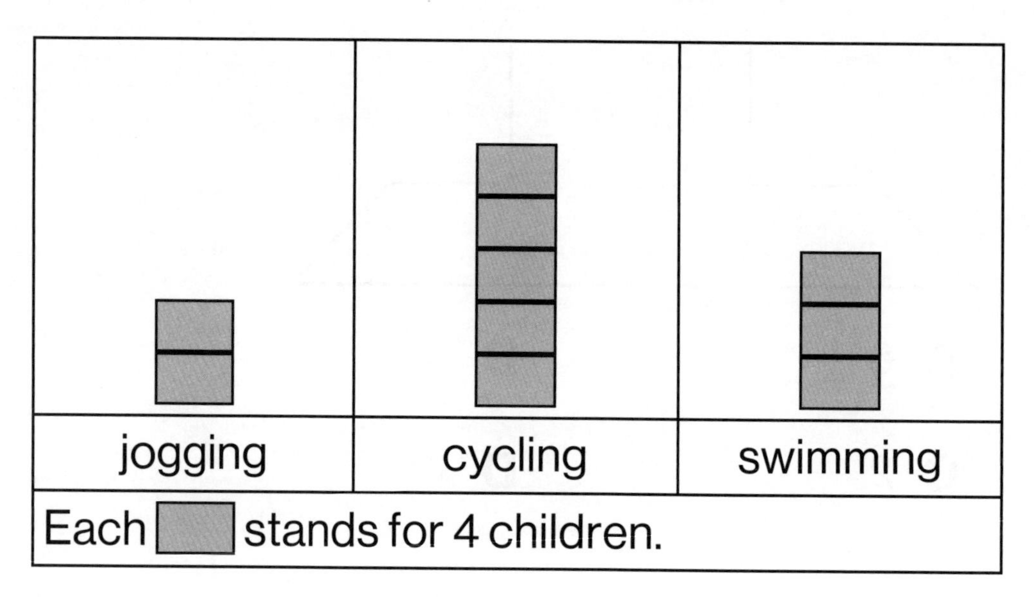

14. How many more students enjoy cycling than jogging?

A 8 **B** 12

C 3 **D** 20

15. How many students are there in Grade 2?

A 10 **B** 20

C 30 **D** 40

Section B (2 points each)

16. Write in words.

17. Arrange the numbers from greatest to smallest.

899 909 809 989

18. Use the digits given to form the smallest 3-digit number.

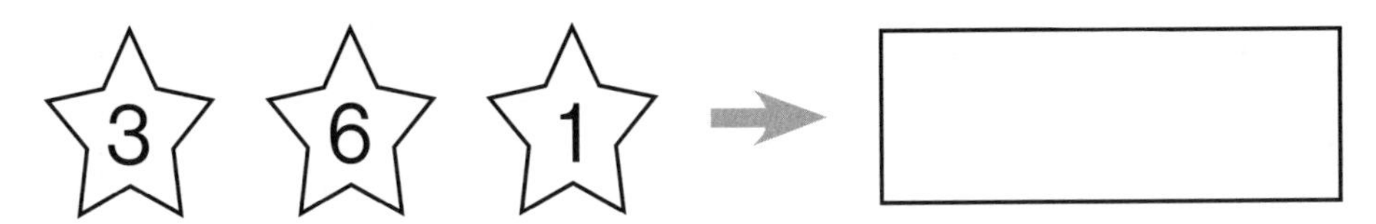

19. The time shown is ______________.

20. The total value of the money below is $ ____________.

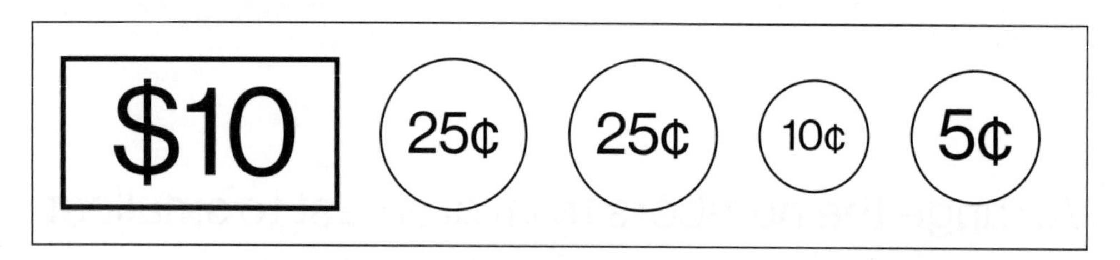

21. Write A, B, C and D in the boxes below using the following clues:

- Pencil A is the longest.
- Pencil A is 2 cm longer than Pencil B.
- Pencil C is not the shortest.

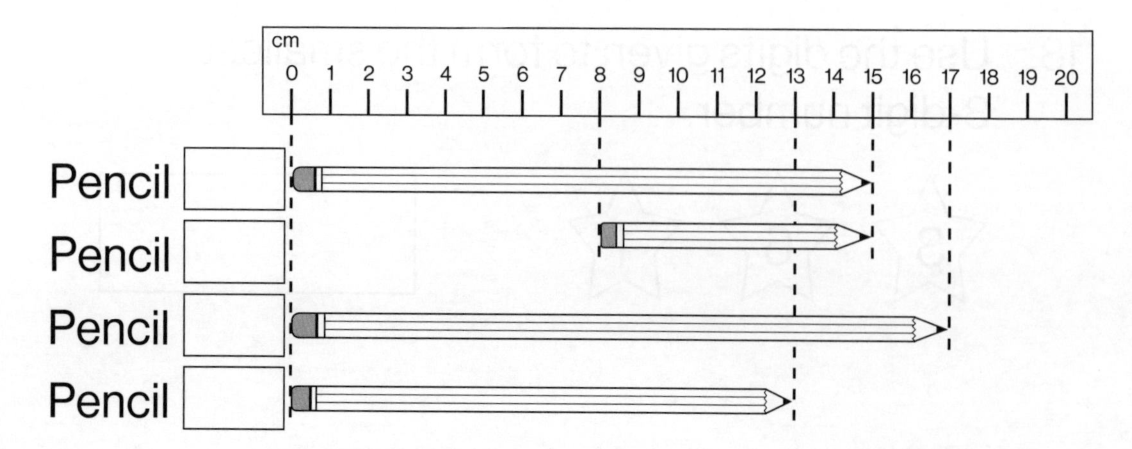

22. Fill in the missing digit.

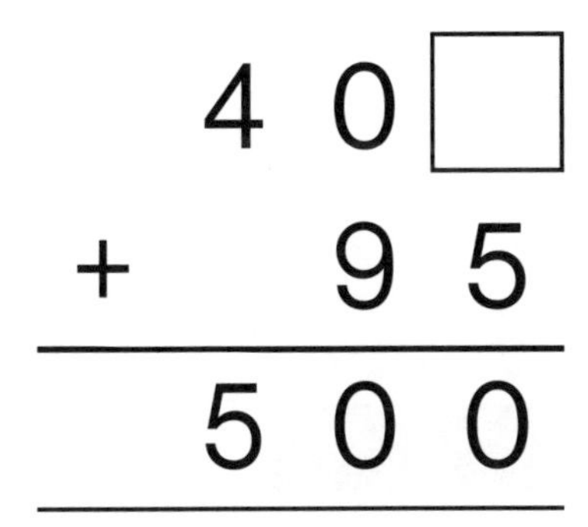

23. The figure has ____________ flat surfaces.

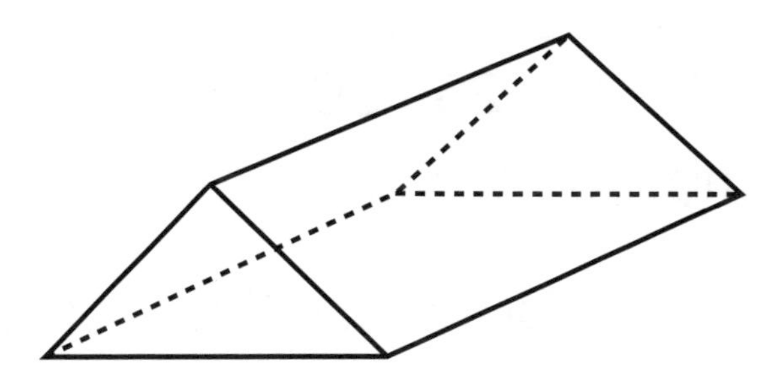

24. What fraction of the figure is shaded?

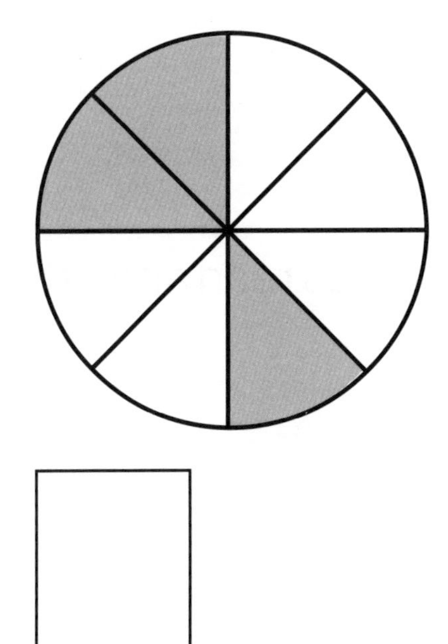

The bar graph shows the number of books 4 friends read over the summer break.
Use it to answer questions 25 and 26.

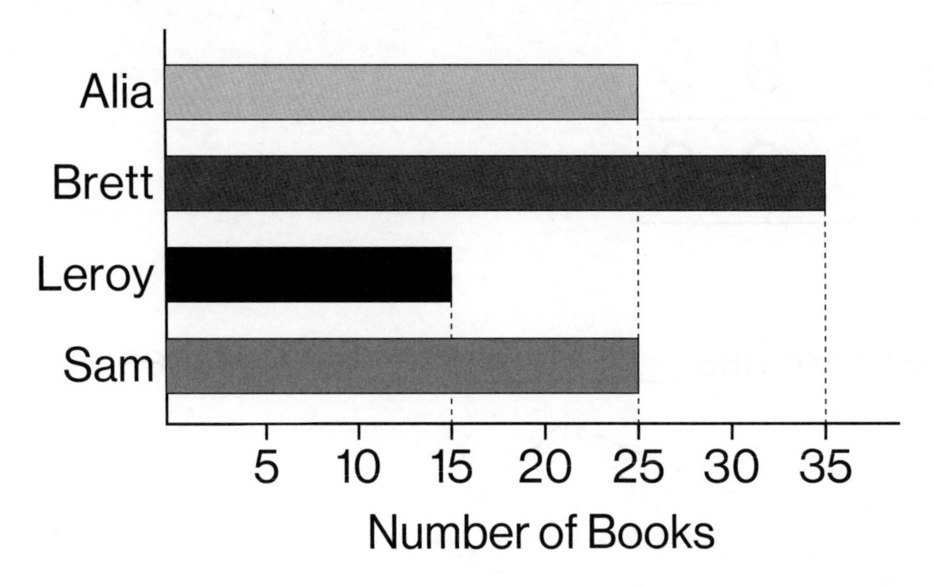

25. How many more books did Brett read than Leroy?

26. How many books did the 4 friends read altogether?

27. Draw 2 straight lines in the figure to make 3 triangles.

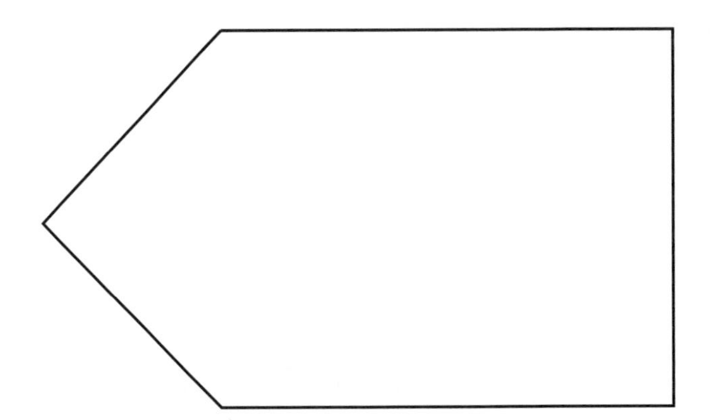

28. Circle A and Circle B are the same size.

2 parts of A = __________ parts of B.

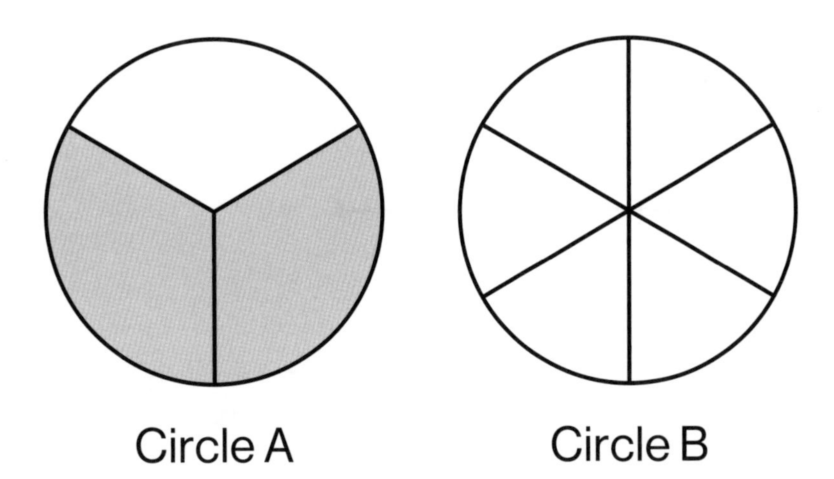

29. Drew cut a string into 2 pieces.
One piece was 140 cm long and the other was 70 cm long.
How long was the string at first?

The string was ☐ cm long at first.

30. Measure the length of each line.
Circle the two lines that have a total length of 9 cm.

A

B

C

D

Section C (4 points each)

31. Susan saved \$2.85.
She saved \$3.15 less than Dylan.

a) How much money did Dylan save?

Dylan saved ____________.

b) How much money did both of them save?

Both of them saved ____________.

32. Malea used 436 paper clips in a project.
Sam used 294 paper clips in the same project.

a) How many more paper clips did Malea use than Sam?

Malea used ______________ more paper clips than Sam.

b) How many paper clips did they both use?

Together they used ______________ paper clips.

33. A 36-cm wire was cut into 4 equal pieces.
Simon joined 3 pieces together.

a) What was the length of each piece of wire?

Each piece of wire was ________________.

b) What was the length of the wire joined together by Simon?

The length of wire joined together by Simon was

________________.

34. Maci has 9 dimes, 4 pennies, 5 nickels, and 9 quarters.
How much money does she have?

Maci has $________________.

35. Tiana's mom cut a cake into 10 equal pieces.
She gave $\frac{2}{10}$ of the cake to Tiana.
How many slices of cake did she give to Tiana?

She gave _____________ slices of cake to Tiana.

Extra Credit

1. Each fruit represents a number.

Answer: ______________________________

2. In a garden, there are 4 lamp posts in a straight row. 5 plants are placed between every 2 lamp posts. How many plants are there altogether?

There are _________________ plants altogether.

BLANK

Answer Key and Detailed Solutions

Unit 6 Test A

1. B
2. C
3. A
4. B
5. D
6. 88
7. 430
8. 50
9. 76
10. 115
11. 11
12. 77
13. 451
14. 203
15. 657

Unit 6 Test B

1. B
2. D
3. A
4. C
5. D
6. 699
7. 22
8. False
9. 899
10. 48
11. 388
12. 105
13. 51
14. 9
15. –

Unit 7 Test A

1. C
2. A
3. B
4. B
5. C
6. There are 3 fish in each group.
7. $8 \times 3 = 24$
   $2 \times 9 = 18$
   $6 \times 5 = 30$
   $4 \times 7 = 28$
   Answer: 6×5
8. 24: 3, 8; 30: 5, 6
9. $5 + 5 + 5 + 5 = \underline{4} \times 5 = \underline{20}$
10. $3 \times 4 = 12$
    $16 \div 4 = 4$
    $5 \times 4 = 20$

Number of tables	1	2	3	4	5
Number of people	4	8	12	16	20

11. $7 \times 5 = 35$
    7 children have 35 bells altogether.
12. $4 \times 5 = 20$
    She bought 20 hair clips.
13. $10 \times 10 = 100$
    There were 100 cupcakes altogether.
14. $30 \div 5 = 6$
    There are 6 chairs in each row.

15a. $6 \times 5 = 30$
There were 30 flowers altogether.

15b. $30 - 8 = 22$
There were 22 flowers left.

Unit 7 Test B

1. B
2. B
3. B
4. B
5. C
6. $6 \times 4 = 4 + 4 + 4 + 4 + \underline{8}$
7. $4 \times 2 = 8$
   Answer: 8
8. $28 \div 4 = 7$
   $16 \div 2 = 8$
   $30 \div 5 = 6$
   $27 \div 3 = 9$
   Answer: $27 \div 3$
9. $\boxed{15} \div \boxed{5} = \boxed{9} \div \boxed{3}$
10. $10 \times 2 = 20$
    $9 \times 3 = 27$
    $7 \times 5 = 35$
    $8 \times 4 = 32$
    Answer: 10×2

11a. $4 + 6 = 10$
There were 10 donuts in each box.

11b. $7 \times 10 = 70$
Francine bought 70 donuts altogether.

12a. $36 \div 4 = 9$
Each student got 9 paper clips.

12b. $9 + 5 = 14$
In the end, Kyla had 14 paper clips.

13. $2 \times 10 = 20$
    $20 + 7 = 27$
    Fiona had 27 stickers in all.
14. $36 - 6 = 30$
    $30 \div 5 = 6$
    There were 6 boxes of eggs.
15. $4 \times 6 = 24$
    $24 \div 3 = 8$
    Each of them got 8 marbles.

Unit 8 Test A

1. C
2. B
3. D
4. B
5. A

6a. $7.05 = 7 dollars 5 cents

6b. $13.40 = 13 dollars 40 cents

7a. Ninety-five cents = $0.95

7b. Twenty-eight dollars and nine cents = $28.09

8. $66.95

9.

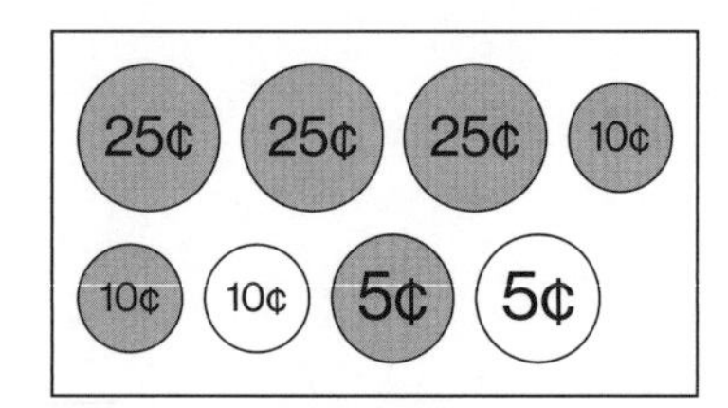

10. 65¢ + 35¢ = 100¢ = $1
Answer: Apple and banana

11. $3.75 + $6.15 = $9.90

12. $8.70 − $5.40 = $3.30

13. $3.20 − $1.30 = $1.90
Answer: $1.90

14. $1.30 + $3.70 = $5
Answer: Lemonade and grapes

15. $5 − $1.30 = $3.70
Answer: $3.70

Unit 8 Test B

1. B
2. B
3. C
4. D
5. A
6. Alyssa
7. 8 quarters
8. $10 $5 $1 25¢ 10¢
$10 $5 $1 10¢ 5¢
$1 $1 $1

9. $1 + $2 + $5 = $8

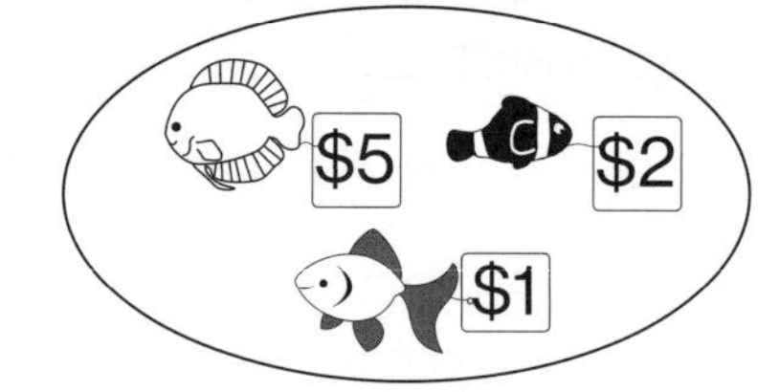

10. $6.30 + $5.95
= $6.30 + $6 − 5¢
= $12.30 − 5¢
= $12.25

11. $6 − $3.95
= $6 − $4 + 5¢
= $2 + 5¢
= $2.05

12. $8.80 − $4.55 = $4.25
Answer: $4.25

13. $7.75 + $2.30 = $10.05
Answer: $10.05

14. $8.80 − $5 = $3.80
Answer: $3.80

15. $78 + $33 = $111
$111 − $45 = $66
Melvin has $66.

Unit 9 Test A

1. C
2. A
3. B
4. D
5. C
6. $\frac{1}{4}$
7. $\frac{7}{8}, \frac{3}{8}, \frac{1}{8}$
8. $\frac{1}{9}, \frac{1}{5}, \frac{1}{2}$
9. 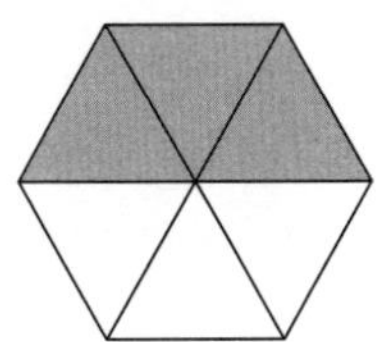
10. $\frac{6}{8}$
11. $\frac{1}{4}$
12. $\frac{3}{8}$
13. $\frac{2}{5}$
14. $\frac{4}{5}$
15. $\frac{7}{7} - \frac{2}{7} = \frac{5}{7}$

    $\frac{5}{7}$ was painted blue.

Unit 9 Test B

1. D
2. C
3. B
4. C
5. D
6. smaller
7. greater
8. 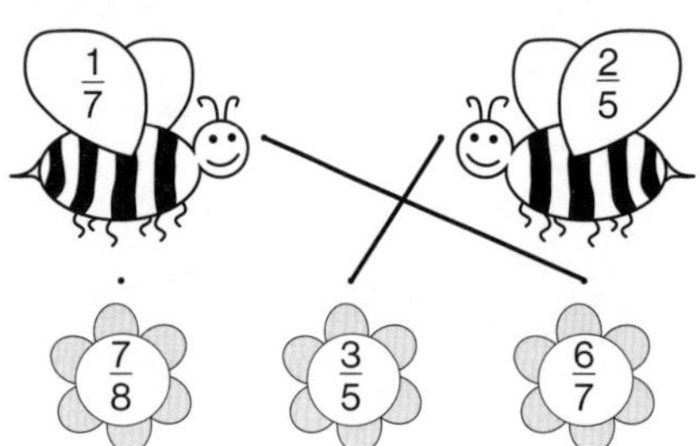

9. $\frac{1}{10}, \frac{1}{5}, \frac{1}{3}, \frac{1}{2}$
10. Draw lines to divide each circle into 4 equal parts.

    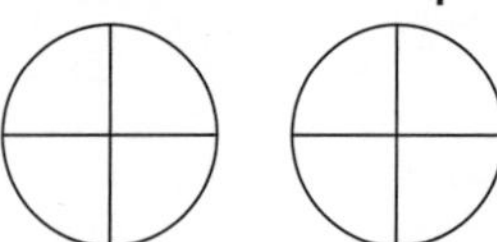

    Answer: 8 quarters
11. Draw lines to divide each square into 2 equal triangles.

    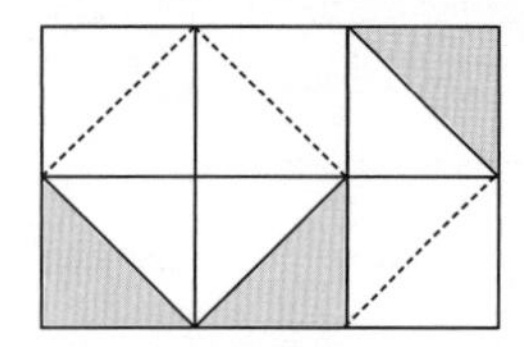

    Answer: $\frac{9}{12}$
12. $\frac{4}{7}$
13. 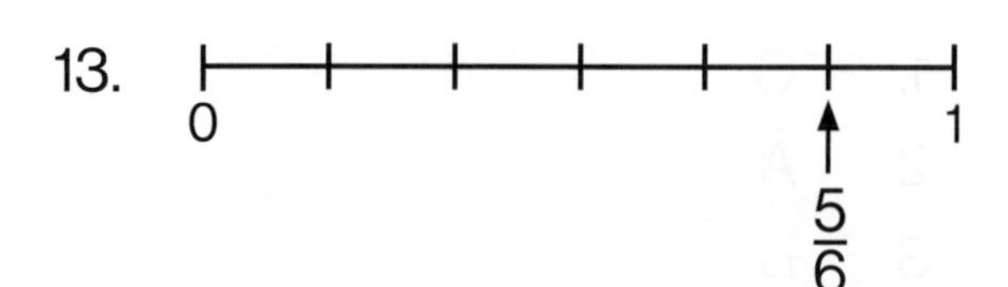

14. $\frac{4}{9}$
15. Pizza B

Continual Assessment 3 Test A

1. B
2. C
3. D
4. B
5. D
6. A
7. C
8. A
9. D
10. C
11. 800
12. Six hundred seventy-nine
13. 525
14. 307
15. 720
16. 0.37
17. 6
18. 6
19. $3 \times 8 = 24$
20. $18 \div 3 = 6$
    Grace can sew lace on 6 dresses.

21a. 75¢ – 50¢ = 25¢

21b. 50¢ + 75¢ = 125¢ = $1.25

22a. Ashton used $\frac{5}{8}$ of the paper.

22b. $1 - \frac{5}{8} = \frac{3}{8}$

There was $\frac{3}{8}$ of the paper left.

23. $260 - 115 = 145$
    She is 145 m from the bus stop.
24. $8.30 – $6.55 = $1.75
    Answer: $1.75
25. $3 \times 10 = 30$ cm
    Answer: 30 cm

Extra Credit

1. The 3 numbers on each line add up to 15.
   $15 - 2 - 9 = 4$
2. $2 \times 6 = 12$
   There are 12 chairs altogether.

Continual Assessment 3 Test B

1. B
2. B
3. C
4. D
5. D
6. D
7. B
8. C
9. D
10. A
11. 881, 818, 188
12. 575
13. 9
14. ×
15. $\frac{1}{5}, \frac{1}{4}, \frac{1}{3}, \frac{1}{2}$
16. $\frac{3}{8}$
17. $0.26
18a. 3
18b. $3 \times 3 = 9$
19. $7 \times 4 = 28$
20. 4
21. $\frac{3}{7}$
22. $2 + 3 = 5$
    $4 \times 5 = 20$
    There are 20 beads in 4 jars.
23a. $1.25 + $2.50 = $3.75
    The scrapbook costs $3.75.
23a. $1.25 + $3.75 = $5
    The folder and the scrapbook cost $5 altogether.
24a. $27 \div 3 = 9$
    She tied 9 presents.
24b. $29 - 27 = 2$
    She had 2 yd left over.
25. 4 × $3 = $12
    $12 − $10 = $2
    Answer: $2

Extra Credit

1. $29 - 15 = 14$
   $14 \div 2 = 7$
2. 2 × $6 = $12
   $12 ÷ 3 = 4
   The cost of each kite is $4.

Unit 10 Test A

1. D
2. C
3. B
4. C
5. B
6. 35 minutes after 2
7. 4:50
8. 20 minutes after 8
9. 6:30
10. 
11. 
12. 
13. 
14. 7:15 P.M.
15. 7:30 A.M.

Unit 10 Test B

1. B
2. B
3. C
4. A
5. D
6. 15
7. 35
8. P.M.
9. 10:10
10. 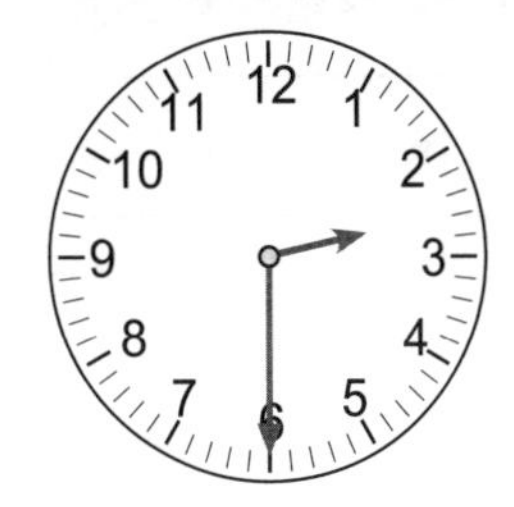
11. 15 minutes
12. 4:55
13. 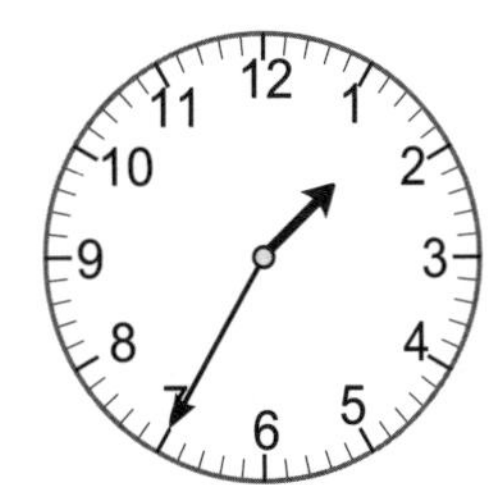
14. 45 minutes after 3 o'clock
15.

Unit 11 Test A

1. D
2. B
3. C
4. C
5. B
6. 8
7. $3 \div 3 = 1$
   $1 + 5 = 6$

goldfish	🐟🐟🐟🐟🐟🐟🐟
swordtail	🐟🐟🐟
guppy	🐟🐟🐟🐟🐟🐟
clownfish	🐟🐟🐟🐟🐟
Each 🐟 stands for 3 fish.	

8. Monday
9. Wednesday and Friday
10. 4 more patients
11. Thursday
12. $12 + 6 = 18$
    Answer: 18 patients
13. 6
14. 13
15. 4

Unit 11 Test B

1. D
2. C
3. A
4. D
5. A
6. $5 + 5 = 10$

   So, = 5

   $5 \times 5 = 25$

   Answer: 25
7. May
8. Ava, Deven, and Tom collected more than 20 stamps.

   Answer: 3 children
9. $5 - 1 = 4$

   $4 \times 10 = 40$

   Answer: 40 more stamps
10. 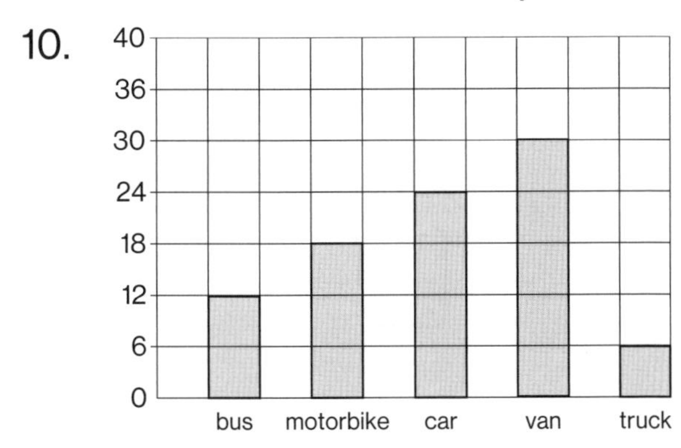

11. $4 + 6 + 8 + 10 + 2 = 30$

    $30 \times 3 = 90$

    Answer: 90 vehicles
12. $10 + 2 = 12$

    $12 \times 3 = 36$

    $36 \times \$5 = \$180$

    Answer: $180
13.

		×			
		×			
		×			
		×	×		
		×	×		
	×	×	×		
	×	×	×		
	×	×	×		×
	×	×	×		×
	×	×	×		×
×	×	×	×	×	×
×	×	×	×	×	×
×	×	×	×	×	×
×	×	×	×	×	×
×	×	×	×	×	×
4	5	6	7	8	9

Height in inches

14. 6 inches
15. 15

Unit 12 Test A

1. C
2. B
3. B
4. D
5. B
6. 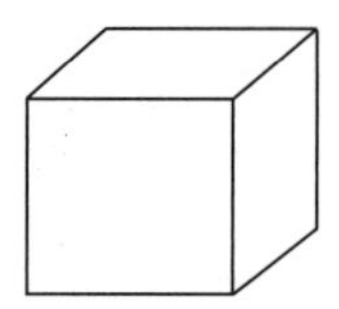
7. 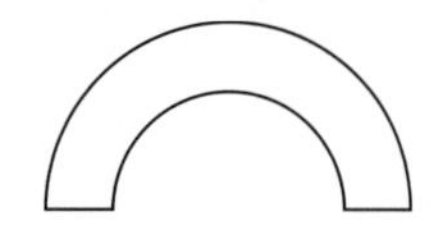
8. triangle, rectangle, half-circle
9. 3 angles
10. 6 quarter circles
11. 2 straight lines, 3 curves
12. 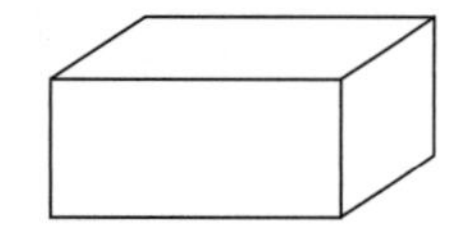
13. 
14. 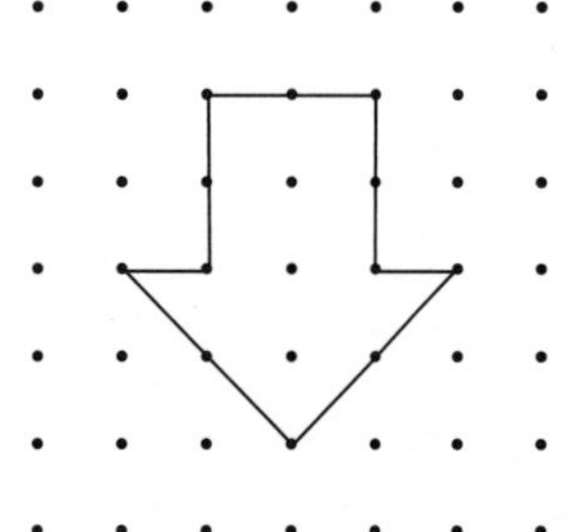
15. 

Unit 12 Test B

1. C
2. B
3. B
4. C
5. B
6. 1 flat surface
7. 6 quarter circles
8. 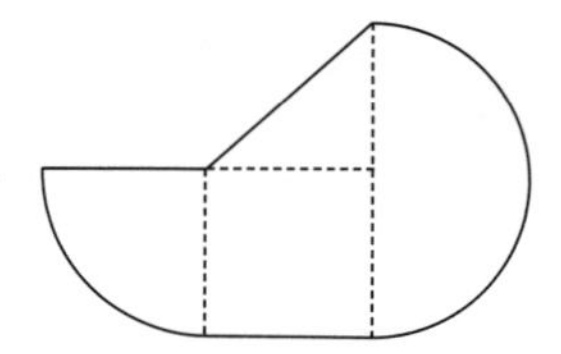
9. 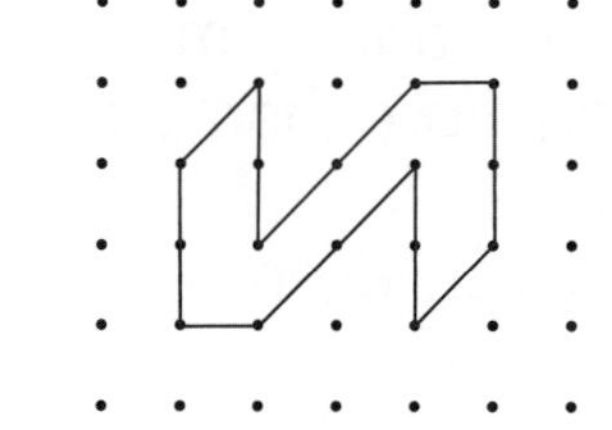
10. 10 angles
11. hexagon
12. 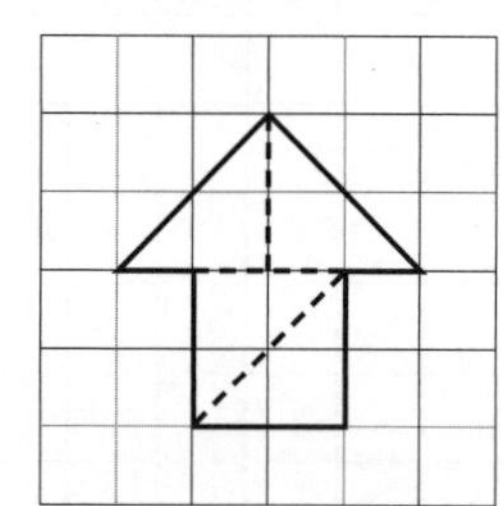
13. 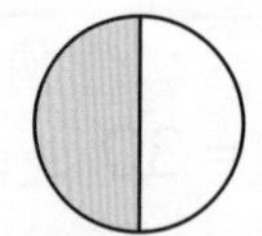
14. 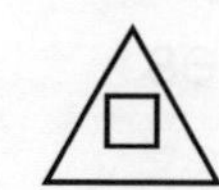
15. 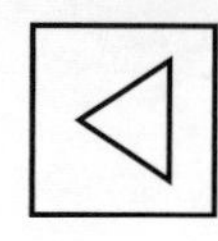

Year-End Assessment Test A

1. D
2. B
3. C
4. A
5. C
6. B
7. A
8. C
9. A
10. D
11. D
12. B
13. C
14. A
15. C
16. five hundred ninety
17. 1:25
18. 70
19. $603 + 106 = 709$
20. $4 \times 6 = 24$
21. 7
22. 3
23. $5 \times 6 = 30$
24. $\frac{1}{11}, \frac{1}{4}, \frac{1}{2}$
25.
26. 5
27. 12
28. B
29. A
30. A and C

31a. $22 - 5 = 17$
Answer: 17 cm

31b. $22 + 17 = 39$
Answer: 39 cm

32a. $2.60 + $3.80 = $6.40
Answer: $6.40

32b. $6.40 + $2.60 = $9
Answer: $9

33a. $35 + 462 = 497$
Answer: 497 apples

33b. $760 - 497 = 263$
Answer: 263 apples

34. $150 + 70 + 20 + 8 = 248$¢
= $2.48
Answer: $2.48

35a. He ate $\frac{3}{8}$ of the cake.

35b. $1 - \frac{3}{8} = \frac{5}{8}$
$\frac{5}{8}$ of the cake was left.

Extra Credit

1. B

2a. $\frac{1}{12}$

2b. $1 - \frac{1}{12} = \frac{11}{12}$
$\frac{11}{12}$ of the pizza was left.

Year-End Assessment Test B

1. C
2. B
3. C
4. C
5. C
6. A
7. A
8. D
9. C
10. D
11. A
12. A
13. C
14. B
15. D
16. eight hundred twelve
17. 989, 909, 899, 809
18. 136
19. 1:55
20. $10.65
21.

B
D
A
C

22. 5
23. 5
24. $\frac{3}{8}$
25. 20
26. 100
27.
28. 4
29. 140 + 70 = 210 cm
Answer: 210 cm
30. A is 4 cm long.
B is 3 cm long.
C is 6 cm long.
D is 2 cm long.
3 + 6 = 9 cm
Answer: B and C

31a. $2.85 + $3.15 = $6
Answer: $6

31b. $2.85 + $6 = $8.85
Answer: $8.85.

32a. 436 − 294 = 142
Answer: 142

32b. 436 + 294 =730
Answer: 730

33a. 36 ÷ 4 = 9 cm
Each piece of wire was 9 cm.

33b. 3 × 9 = 27
The length of wire joined together by Simon was 27 cm.

34. 90¢ + 4¢ + 20¢ + 225¢ = 339¢ = $3.39
Answer: $3.39
35. 2

Extra Credit

1. 3 × 3 = 9
So, (strawberry) = 3
18 ÷ 3 = 6
So, (apple) = 18
Answer: 18
2. ▯ooooo▯ooooo▯ooooo▯
3 × 5 = 15
There are 15 plants altogether.